DIRT TRACK
CHASSIS & SUSPENSION

ADVANCED SETUP AND DESIGN TECHNOLOGY

FOR DIRT TRACK RACING

FROM THE EDITORS OF
CIRCLE TRACK MAGAZINE

HPBOOKS

HPBooks

Published by the Penguin Group
Penguin Group (USA) Inc.
375 Hudson Street, New York, New York 10014, USA
Penguin Group (Canada), 90 Eglinton Avenue East, Suite 700, Toronto, Ontario M4P 2Y3, Canada
(a division of Pearson Penguin Canada Inc.)
Penguin Books Ltd., 80 Strand, London WC2R 0RL, England
Penguin Group Ireland, 25 St. Stephen's Green, Dublin 2, Ireland (a division of Penguin Books Ltd.)
Penguin Group (Australia), 250 Camberwell Road, Camberwell, Victoria 3124, Australia
(a division of Pearson Australia Group Pty. Ltd.)
Penguin Books India Pvt. Ltd., 11 Community Centre, Panchsheel Park, New Delhi—110 017, India
Penguin Group (NZ), 67 Apollo Drive, Rosedale, North Shore 0745, Auckland, New Zealand
(a division of Pearson New Zealand Ltd.)
Penguin Books (South Africa) (Pty.) Ltd., 24 Sturdee Avenue, Rosebank, Johannesburg 2196, South Africa

Penguin Books Ltd., Registered Offices: 80 Strand, London WC2R 0RL, England

While the author has made every effort to provide accurate telephone numbers and Internet addresses at the time of publication, neither the publisher nor the author assumes any responsibility for errors, or for changes that occur after publication. Further, publisher does not have any control over and does not assume any responsibility for author or third-party websites or their content.

DIRT TRACK CHASSIS & SUSPENSION

First edition: July 2007

ISBN: 978-1-55788-511-1

PRINTED IN THE UNITED STATES OF AMERICA

10th Printing

NOTICE: The information in this book is true and complete to the best of our knowledge. All recommendations on parts and procedures are made without any guarantees on the part of the author or the publisher. Tampering with, altering, modifying, or removing any emissions-control device is a violation of federal law. Author and publisher disclaim all liability incurred in connection with the use of this information.

CONTENTS

ACKNOWLEDGMENTS

Special thanks to Bob Bolles, Scott Bloomquist, Tom Hintz, Billy Moyer, Bob Ryder, Jim Doffing, John Clark, Donald Nosek, Michael Thomas, Bob Ryder, Will Handzel, Jon Fitzsimmons, Tom Rounds, Ronnie Johnson, Nick Masters, Tony Hammett. Also thanks to Sean Holzman and Rick Stark of Primedia Enterprises for making this book possible.

Circle Track magazine is the number one source for advanced technical racing information. It has been the leader in presenting state-of-the-art technical information for the racing community for over twenty-five years and now offers a book featuring some of its best dirt track racing technical articles. *Dirt Track Chassis & Suspension* contains valuable information about how to set up and race your dirt car.

Dirt track racing, from a technical perspective, is one of the most difficult forms of motorsports. Many of the classes of dirt late model and modified race cars have a complexity of design that exceeds even the much more expensive and "refined" race cars in Nextel Cup, IndyCar or Formula One.

It is this difficulty of designing and setting up the dirt cars that interests the technical staff at *Circle Track* magazine. A significant portion of our readership is involved in dirt track racing, so it is with a great deal of pride that we are able to provide assistance and guidance to these teams

Some of the most knowledgeable writers in the racing industry have contributed to the pages of *Circle Track* over the years and this book puts a great deal of valuable information at your fingertips. It will be a worthwhile addition to your racing technical library. Much of the information contained in this book is timeless and all of the articles have valuable insight into the subject matter.

On the following pages, you'll find information from racers and technicians who make their living designing, setting up and writing about dirt race cars. Racers Billy Moyer and Scott Bloomquist; former CT technical editors Tom Hintz, Will Handzel, Jon Fitzsimmons and Tom Rounds; and myself, the current technical editor, offer some of the best insights into the mysterious world of dirt car chassis and suspension set up in the sport.

Even though some of the information in this book applies to a particular type of dirt track racing, you'll be able to learn from the technology and apply it to your own car. Dirt chassis engineering is all interrelated to a certain extent. We hope that this information will make your racing effort more successful and, more importantly, more enjoyable. If so, all of our efforts of the past will have been worth it. Thanks for being such loyal readers and supporters of *Circle Track* magazine.

—*Bob Bolles, Senior Technical Editor*

BASIC DIRT CHASSIS
THEORY & DESIGN

Chapter 1
DIRT CHASSIS BASICS

What Makes an IMCA Dirt Modified Work?

by Bob Ryder and Jim Doffing

Pre-race preparation and understanding chassis setup basics is always the major priority of any successful race team. How you prep your car before each race will tend to show at the pay window and in the point standings at the end of the season. You want to be able to roll your car off the trailer race-ready, only having to make minor chassis adjustments.

Jim Doffing, owner of Flexi-Flyer, invited us over for a track test with a Flexi-Flyer Dirt Modified. Terry Belcher Jr., a 16-year-old rookie sensation, would be our throttle robot during the track test. As we talked with Jim about all the different things that make these cars so unique, he couldn't stress enough about pre-race preparation and setup. Just as with all other successful racing teams, it all comes from thorough preparation and correct chassis setup.

We wanted to learn more about these cars and what it takes to make them perform to their full potential. Jim was kind enough to walk us through the proper pre-race chassis setup. He told us the major problem with these IMCA Modifieds is that they are an underpowered race car, with a low power-to-weight ratio. He also stated that chassis setup is very critical.

A more high-powered car has the torque and horsepower to overcome some handling ills, but the 250 to 300 fewer horsepower in an IMCA Modified makes correct chassis setup extremely critical. Everything must be in sync, because you can't afford to have front tires that point one way and the car the other, or a rearend that tracks at an angle (crabbin') down the straightaways. Camber, toe, and steering all have to be spot-on, and the rearend has to be square to the chassis as the body rolls in order to keep the tires' contact patches from scrubbing off at speed.

CHASSIS BINDS

The first thing you should do before any chassis setup is performed in the shop, is to look for the presence of any chassis binds in the car. You need to move each of the wheels through at least 2 inches more than their normal wheel travel. Observe the movement of everything attached to that wheel. Look for shock absorber binding or bottoming out, A-arms moving freely or contacting the frame, the steering shaft moving freely when turned without contacting anything, free movement of all steering components through the full range of left-to-right steering with no binding or contacting, the Panhard bar moving freely with no binds or without contacting any chassis parts, and the rear suspension arms moving freely with no binds. At this point, we didn't see any problems with binding of suspension components.

SQUARING THE REAREND

When making the rearend square you need to make sure the rearend housing is set straight in the car, perpendicular to the vehicle centerline and not angled. If the right-rear tire is set behind the left-rear tire, the car will be too loose. If the right rear is set ahead of the left rear, it will push.

Squaring the rearend is very critical to the car's handling. Even being 0.25 inch out of square can have a significant effect on handling.

Most chassis manufacturers have built-in squaring reference marks or holes in the framerails. Using these reference marks, measure straight back to the rearend housing at the same point on each side of the reference point. One of the most common squaring procedures is called "stringing" the car. It is done by stretching string tightly between jack stands that are set up on either side of

Terry Belcher Jr.'s Flexi-Flyer runs on alcohol, which burns almost twice as fast as gasoline. The 31-gallon fuel cell is always filled to its max. (Since dirt requires more rear weight, this is easier to do with fuel.) With the larger volume of fuel involved there is always a big difference in the handling characteristics of the car from the beginning, middle, and end of the race, because as the rearend becomes lighter as the race goes on, it will naturally become looser.

After a few laps of wheel packing and feeling the chassis out, the revs come up and the Challenger Racing 358ci Chevrolet small-block comes to life. Terry pitches it sideways, because on a dirt car you want a higher roll center than that of an asphalt car. The reason for the higher roll center is to get more body roll in order to transfer the weight to the outside, which helps give you right-rear side tire bite to accelerate off the corners faster.

the car and which run parallel to the wheelbase to establish outside reference lines. Measure from the string at several points to the framerails to make sure the string reference line is absolutely parallel to the rails and that each string is parallel to the other. Then measure from the string to the front side and the rear side of each rear tire. Make sure the measurement hits the rear tire in the same point on each side. These measurements should be identical. If they are not, the rearend housing is not square.

If the front measurement on the right side is greater than the rear one, it indicates that the rearend is pulled ahead on the right side. Lengthen the right-side trailing link slightly and remeasure. If a string is stretched between the two rear jack stands running parallel to the rearend housing, measurements can be made from it forward to the rearend housing on each side to double-check the square.

SETTING RIDE-HEIGHT

Ride-height is always set before the weight distribution is set. Ride-height is measured from the flat of the bottom of the framerails at the forward and rearward corners of each side of each rail. Once ride-height is set, the front/rear and left/right weight distribution can be set on scales.

For dirt applications we use the following corner heights:

Left-front tire: 4.5 inches
Right-front tire: 5 inches

Left-rear tire: 4.75 inches
Right-rear tire: 5.5 inches

Dirt track cars need more ride-height (ground clearance) than asphalt cars to improve side bite (caused by more overturning movement and body roll). For cars running on dirt tracks that get very hard, dry, and slick, the ride-height should be the same as on asphalt cars. For very loose and wet tracks, the ride-height needs to be slightly higher than the numbers quoted here to help you get more side bite.

BRAKES

One of the most troublesome areas that is often neglected is brake bias. If you have a push or a loose condition as you enter the corner, the first thing you should suspect is brake proportioning.

1. Too much rear brake causes a car to be loose.
2. Too much front brake causes a push.

On a dirt application, more rear bias is needed, usually somewhere between 60 to 65 percent.

Jim Doffing checks tire stagger and records the numbers before testing to get a track-side baseline. Different track conditions can be altered by changing stagger. If you learn to understand stagger you will save a lot of time and aggravation in fine-tuning a chassis.

During the session, Terry Belcher Jr. starts by driving the high groove and works his way down to the bottom, because the top tends to dry out quicker as the moisture goes to the inside of the corners. And this track dried out quickly, since we did this test in August in the middle of the Arizona desert, where the ambient temperature was 118 degrees F on the track.

PANHARD BAR

The Panhard bar is the simplest and most widely used of the lateral-control linkage systems. The bar is attached to the chassis at one side and attached to the axle housing on the opposite side. The Panhard bar is a long tube with spherical rod end bearings at each end. This tube braces one side of the chassis against the other, resulting in no side-shift of the axle.

Cars that use a short Panhard bar will have it mounted on the left side of the chassis for dirt use. Mounting the Panhard bar on the left side offsets the roll center roughly halfway between the vehicle

center line and the left-rear tire center. Also, mounting the Panhard bar on the left side creates a rear roll center that decreases as the chassis rolls. It will create more body roll, which makes for more bite on the right-rear tire, thus getting you off the corner quicker. When setting a Panhard bar, you should make sure it is level and set after the chassis setup has been completed. For dirt, you'll want to set it at 11.50 inches from level ground.

REAR TIRE STAGGER

What is stagger? Stagger is the difference in inches of the tire circumference between the left-rear and right-rear tires. When the right-rear tire is larger in circumference than the left-rear you have stagger. When the left-rear is larger in circumference than the right-rear, you have reverse stagger.

Stagger helps get the car into the corner without pushing. It will also help overcome the pushing tendency of a car as it changes direction from straight-line running.

Tire stagger will get a car off the corners more quickly, which means more straightaway speed. Improved corner speed with more straightaway speed helps the driver set up and pass more easily. Does stagger hinder straightaway speed? Yes, because it's a dragging force. But more acceleration off a corner more than compensates for this. You need stagger to launch the car off the corner to gain more straight-line speed.

The minimum stagger required will vary from car to car and track to track. The variables include car track width (left-rear tire center line to right-rear tire center line) and the turn radius of the racetrack. Minimum stagger is the difference in size of the right-rear tire from the left-rear tire, determined by these variables. Because the two tires are running on different radii, one must travel farther than the other (the outside tire must travel a farther distance on a wider arc). This is accomplished by the outside tire being larger in diameter than the inside tire so that it runs at a slightly faster speed.

The tighter the corner, the more stagger the car needs. If your car is pushing in the middle of the corner and beyond, more stagger is required. If the car is loose, decrease the stagger.

Street-type IMCA tires require more stagger, but it is harder to come by than with regular racing tires. Because of the design of the tire, street-type tires do not react very well to circumference changes with air pressure increases. It is difficult to get much stagger difference out of these tires. The best thing to do is measure different tires to find as much stagger as possible.

Also, you need to remember that with rear stagger

Jim Doffing checks the right-front stagger with a—what else?—stagger gauge. By changing the front stagger you can change the car's turn-in characteristics.

If you have done your homework and set the car up correctly at the shop, once you're at the track you will not have to make many changes. You can fine-tune your chassis setup by just changing the stagger and wheel offset. The more testing is done, the better you will know the true parameters of the driver/car combination.

The secret to success for any race team is seat time. The more seat time the driver is able to acquire, the better his or her skills become as a driver and communicator to his crew. The driver is the middle man of the team (car-driver-crew), and all three need to be interlinked before the team can become a winner.

comes rear camber on a straight axle. If we put a smaller circumference tire on the left-rear, it puts negative camber in the right-rear and positive camber in the left-rear. This will help put the tire footprint flat on the track during cornering.

Front stagger is also used on these cars, but it does not work like rear stagger, simply because the front wheels are not connected to the same axle. Front stagger is merely a tuning adjustment which adjusts corner height (giving you weight jacking).

The ultimate guideline for rear stagger is to use as much as is necessary to get the car to roll easily around the middle of the corner and not get loose during a corner exit. Tire temperature is also monitored closely to be sure it stays within reason across the right-rear (the inside edge of the right-rear is going to be hotter, showing a tire temperature spread very similar to the right-front with proper camber).

The secret to going fast at any track, dirt or asphalt, is to keep the momentum up through the middle of the corner. We are running on two different radii through a corner, and stagger is what will make a car run on the different radii. This will keep the car freed up so it will roll around a corner instead of having to be "driven" around the corner. This will allow you to pick up the throttle without getting the rearend loose or creating a push.

Stagger versus crossweight is also a very important consideration. A car with a high amount of crossweight is affected more by stagger than a car with less crossweight. A car with a very heavily loaded left-rear wheel is more critical for stagger.

FRONT TIRE STAGGER

When the front suspension is set for proper ride-height and geometry, the last thing you want to do

is use the front weight jack bolts to jack weight into the chassis. This will alter the front end and steering geometry. A better way to jack weight at the front is with front stagger. One inch of front stagger is a common starting point. Front stagger adds wedge (or crossweight) to the chassis because the right-front corner is higher than the left front. It puts tilt in the chassis at the front, which adds weight at the right-front and left-rear, and subtracts weight from the left front and right-rear. Front stagger can be increased to add more wedge, or decreased to take some weight out of the left-rear.

10 BASIC GUIDELINES
FOR USING STAGGER

1. More stagger makes a car looser coming off the corner and heading into the corner.

2. Less stagger tightens up the rear of the car.

3. If a car pushes coming off the corner, add more stagger.

4. The less rear weight percentage a car has, the less stagger the car needs.

5. The tighter the turn (shorter turn radius), the more stagger is required. For example, the minimum stagger required with a 64-inch rear track on a tight quarter-mile track might be 2 inches, but the same car on a wide sweeping 3⁄4-mile track (with the same banking) might be only 1⁄2 inch.

6. A wet, heavy dirt track will use much more stagger to help turn the car: somewhere between 2.5 to 3 inches with a 64-inch-rear track car on a 3⁄8-mile track.

7. A dry, slick track will use much less stagger. For example, with the same car and track as noted in the above example, the dry, slick track would only require 0 to 1⁄2 inch of stagger.

8. More crossweight requires more stagger to get the car into and out of the corners. Conversely, less crossweight requires less stagger.

9. When picking out unmounted tires, keep in mind that the tires will grow about 0.5 to 0.75 inch after they are mounted.

10. To get a good indication of how a tire will grow in circumference when run, mount the tire and increase the right-side tires to 40 psi and hold them there for 20 minutes (first be sure that it is safe to increase the tire pressure that high). Then bleed the tires down to 20 psi and immediately take a circumference measurement. That number should give you a good indication of the final tire dimensions after they are run. Mark the sidewalls with the circumference measurement. Different bias-ply racing tires will grow at different rates, even though they are theoretically produced equally.

Manufacturer
Flexi-Flyer
P.O. Box 34641
Phoenix, AZ 85067
Orders: (800) 528-2059
Tech: (602) 264-7646

Chapter 2

How to Install and Tune Leaf Springs

Some Stock car classes make it to your advantage to switch to leaf springs.

Text and Photos by John Clark

Leaf springs have been around since the covered wagon days when they were used as driver-suspension devices. While the technology behind multileaf springs has improved vastly, these springs are often looked down on by racers. In fact, they possess many desirable suspension features such as roll oversteer, high lateral stiffness, cushioned forward-bite, and high anti-squat percentage; the springs also make chassis setups more forgiving. For Stock cars that race on dirt, the rugged simplicity of a multileaf rear suspension can be a good, and fast, choice.

The leaf rear suspension consists of two springs, a couple of bushings and some sort of shackle. Nice and simple. The Chevrolet Camaro is the most popular of leaf-sprung Stock cars, but later year models, which have unusable strut front ends, limited the numbers and availability of parts. Many local dirt tracks have a phenomenon that is sometimes known as the "Camaro Parade" since many of the top cars are leaf-spring Camaros. Some tracks allow switching other kinds of cars to leaf springs to promote equality among manufacturers and to reduce the cost of racing.

One such track is the Southern New Mexico Speedway in Las Cruces, New Mexico. The Stock car division has a field of 40 to 50 full-blown Stock cars competing with a minimum of rules, yet the track features very tight racing at

a yearly cost (including motor) less than an IMCA Modified. These cars weigh 3200 pounds, sport stock frames or front clips, and run unmodified two-barrel carburetors, track tires, and stock-type suspension components to control costs. Yet, to allow any type or make of car to be competitive, the rear suspension rules basically say, "Run it like it was produced or switch to leaf springs." More than three-fourths of the class, including all the front runners use leaf springs.

The installation of leaf springs is simple and straight-forward as are the parts used. To find out more about leaf-spring suspensions, we talked with the folks at Landrum Springs and AFCO Products and thank them for their patience and the information they provided.

SPRING TYPES

There are three types of leaf springs: monoleafs, Chrysler, and Camaro. Monoleafs are a single master leaf used on Late Model–type race cars. Monoleafs require secondary coil springs to actually hold the weight of the car, and therefore complicate the rear suspension.

Multileaf springs come in two principle variations, Chrysler and Camaro. Multileaf springs are modeled after their production counterparts. Landrum Spring offers six

Front leaf-spring mounts should have several mounting heights. By using various combinations of holes different ride-heights and anti-squat percentages can be experimented with. Remember, the front mounts are subjected to the entire forward force from the rear axle, which makes stout mounting and gussets necessary.

An adjustable lowering block should be used to allow the rear axle to be squared to the chassis centerline. The adjustable block can be used to tune for various track conditions. A right-side wheelbase longer than the left loosens the car coming out. A right-side wheelbase shorter than the left tightens the car coming out. Try 1/4-inch increments. The U-bolts on both sides should be loosened before adjusting the block.

rates for Camaro and 10 rates for Chrysler geometry, which is the more popular of the two. This is because Chrysler uses an asymmetric design in which the short, stiff front section behaves like a cushioned traction bar on acceleration. The long, supple tail section is the "spring" portion which supports the weight of the car. A Camaro spring is more symmetrical front to rear, and the shorter tail section tries to counteract the traction bar action of the front half. If the rules allow, Chrysler springs

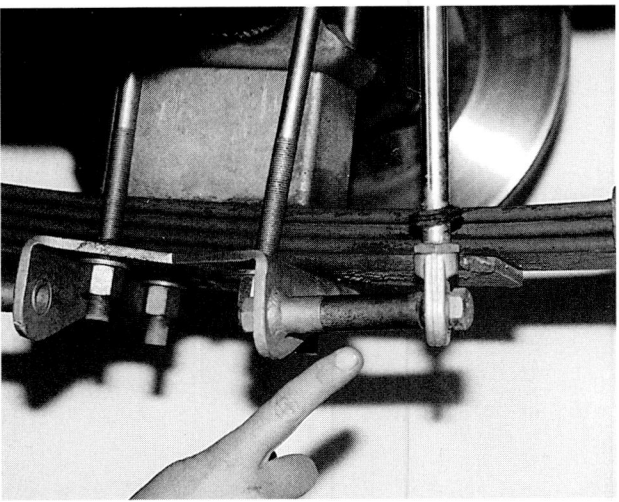

If your rules don't allow a 90-10 damper shock to be mounted above the rearend housing, try spacing the shock 2 to 3 inches out from the lower mount plate. The left shock should be in front of the rear axle and the right mounted to the rear. This uses the shocks to help damp the rising and falling of the pinion during acceleration and braking. Locating the left shock in front of the axle plate loads the left-rear tire first.

appear to be the way to go, although using Camaro geometry springs is not all bad. The aftermarket racing-spring companies are stiffening the front and softening the rear section to bring the Camaro geometry much closer to the Chrysler in performance.

FRONT BUSHINGS

Stock-rubber front bushings should have shoulders to prevent side-loads from sliding the bushing inside the spring eye. If stock-rubber bushings are used, hard plastic washers make ideal shims to fill the gap between the spring bushing and the mounting box. Stock Chrysler front spring bushings are of a small enough diameter to prevent excessive deflections. The larger diameter GM or Camaro rubber bushings can deflect sideways and upward, allowing the differential to move erratically, degrading the handling.

Front- and rear-spring bushings should be torsionally stiff over the long axis of the spring. Torsional rigidity increases chassis rear roll stiffness without significantly increasing rear-wheel bump rates. The stiff bushings help twist the leaf spring to approximate a "poor man's" rear antiroll bar. Stock-rubber bushings do not have good torsional rigidity. Hard plastic or urethane are better, but aluminum bushings with steel center sleeves are the stiffest. AFCO sells a spherical bearing front "bushing" assembly that eliminates the twisting on the front half of the spring, freeing up the spring during body roll.

Slider blocks should be checked for smooth, bind-free operation. This slider was damaged by minor wheel contact with another car. Notice how the side of the slider is bent away from the spring eye. Subsequent checking also revealed a broken bearing. Minor bends can be straightened with a large crescent wrench, but it is wise to carry a small parts set to the track or even a spare block assembly.

Ball-bearing slider blocks are considered "the Cadillac" of rear spring mounting arrangements. Block-type sliders are just as effective, but may require occasional cleaning and greasing. A dry moly-disulfide type lube is best since no sticky residue is left. Slider blocks can be mounted either with bolts or tack-welded in place. To complement the multiple-hold front spring mounts, the rear spring mounts should also be height adjustable.

AXLE MOUNTING

The usual method for mounting a rear-end housing to leaf springs is by using some type of spring pad, a lowering block, and lower spring plates clamped together with two long U-bolts. There are two basic types of spring pads: clamp-on and weld-on. Clamp-on pads are more convenient and are often used on more expensive full-floating rearends. On heavy cars clamp-on pads have been known to slip and often require tack welding to

Notice the 10-inch height difference between the front and rear mounting points of the leaf spring as installed at the suggested starting point. Reducing this height difference increases the anti-squat or forward-bite. Remember that leaf springs have a very high anti-squat percentage. If the anti-squat is raised over 100 percent, no more forward-bite is produced, the rear of the chassis will only rise on acceleration. High anti-squat will cause wheel hop or extreme looseness when the brakes are applied, especially if 90–10 damper shock(s) cannot be used.

hold them. Remember, there is as much as 3000 lb-ft of torque that is trying to rip off the pads. Weld-on axle pads are much less expensive and have no slipping problem.

When you are installing either type of pad, be sure both are flat and square to each other and to the leaf-spring surface. With both U-bolts loose the spring pads should sit flat on the leaf springs with no twisting or front-to-back preloads for the best performance. Weld-on pads should be welded about 1 inch at a time to allow the axle housing to cool between welds, to help minimize heat-induced housing distortion.

LOWERING BLOCKS

Lowering blocks are used for two reasons: 1) to lower the rear of the chassis; 2) to lower the rear roll center. Blocks are available in 1/2-inch steps up to 4 inches, but blocks more than 3 inches long can cause erratic spring-loading. Lowering blocks can be angle-milled to correct the pinion angle (usually 4 degrees down). Special lowering blocks are also available with an adjustment that allows the center bolt to move an inch to the front or rear for squaring the rearend to the chassis center line (or deliberately setting it out of square). The best starting point is to square the rear-end housing to the chassis center line.

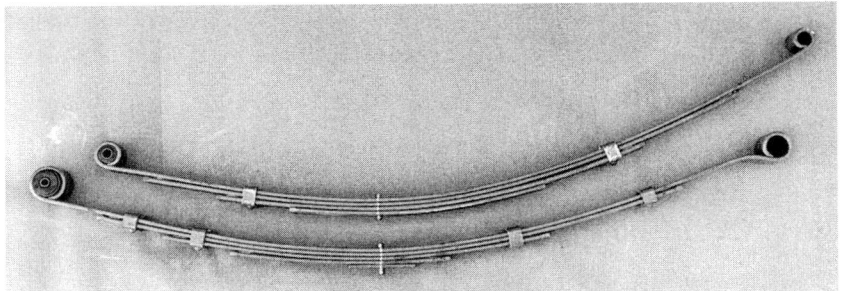

The difference between a Camaro-type spring (lower) and a Chrysler-type spring (upper) is noticeable when placed side by side. Notice the shorter, stiffer front section and long, slender tail section (right side) of the Chrysler. The Chrysler spring allows the front section to behave as a short traction bar while the tail supports the weight. Camaro springs are not the first choice, but are quite capable of being competitive.

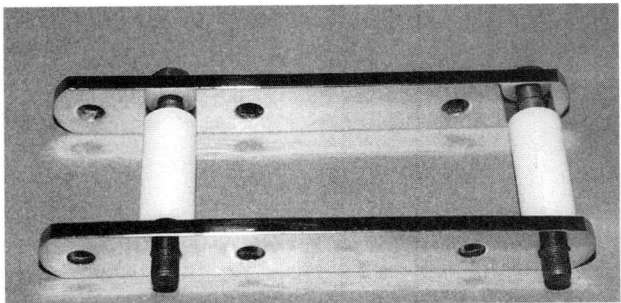

Flat plate- and bolt-type shackles may work on the street but have no place on an oval-track race car. The lack of lateral rigidity will allow the rear axle to shift from side to side during cornering.

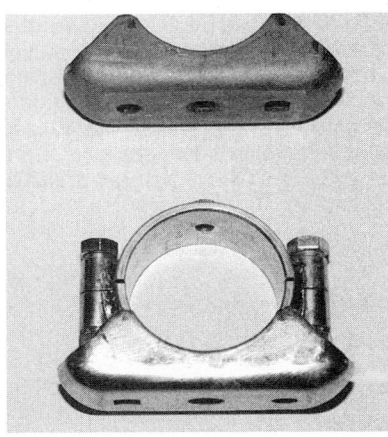

On the top is a weld-on axle pad, the lower is a clamp-on style. Clamp-on mounts are designed to be used on aftermarket-type rearends using 3-inch o.d. tubes. Factory housings are not usually the correct size and often need weld-on pads. Clamp mounts should be checked for slipping and tack-welded if necessary.

U-BOLTS AND MOUNT PLATES

U-bolts used for racing purposes need to be made of Grade 5 (or equivalent) material. The recommended torque for 1/2-inch fine-thread U-bolts is 50 to 55 lb-ft using double-height nuts and very thick hardened washers. U-bolts should be re-torqued after the first night and then every three to four races. The lower mounting plates must be thick and have flanges to resist bending. Mounting plates are available with either right or left shock mounts or in a universal-type that can be installed on either side.

SHACKLES OR SLIDERS

The rear of the leaf spring is usually mounted with either conventional-type swinging shackles or a slider-block device. If conventional shackles are used they must be able to carry a side-load without deflecting. Street-type shackle kits which consist of two flat metal straps with holes are simply not rigid enough for oval-track racing. Strap-type shackles can be reinforced with either a lengthwise flat plate or a piece of square box tubing that is welded between the straps. Shackles that are shorter than 3 inches should be avoided to eliminate the possibility of over-centering. Conventional swinging shackles exhibit a slight weight-jacking effect as they move through their arc.

A more consistent and rigid mounting for the rear of the spring is a slider block, which is available with either a roller-bearing or sliding nylon block.

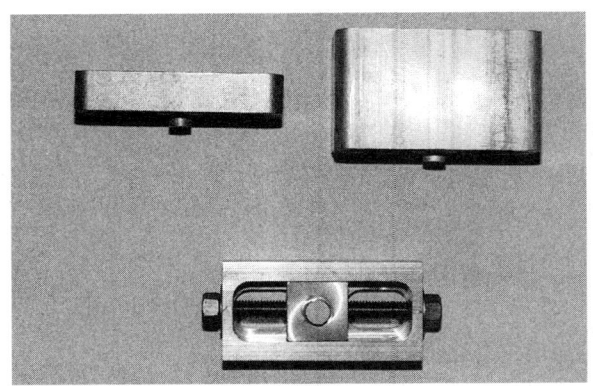

Lowering blocks come in different heights and materials but blocks more than 3 inches in height can cause erratic handling. The lower block is an adjustable Coleman Machine part designed to allow adjusting the rear axle alignment/wheelbase on individual sides.

Roller-bearing blocks have the least friction but are more susceptible to damage from racing contact. Slider blocks of either design should be periodically cleaned and checked for smooth, bind-free operation. It would be wise to carry a spare set of small parts to repair slider blocks at the racetrack. Slider blocks are also much easier to mount high in the rear of the chassis for proper spring geometry and have much less tendency to bind under high side-loads. Sliders should be mounted so the slot "points" to the front eye of the spring to prevent increasing or decreasing the spring rate as the slide moves under load.

TUNING

Leaf springs have some built-in roll-steer: the right side lengthens, and the left side shortens under body roll to the right. This makes the car looser as the body rolls, good for entering the turn. Increasing the arch in the front section of the spring adds more roll-steer as does lowering the front eye of the spring. Moving the front eyes of the springs

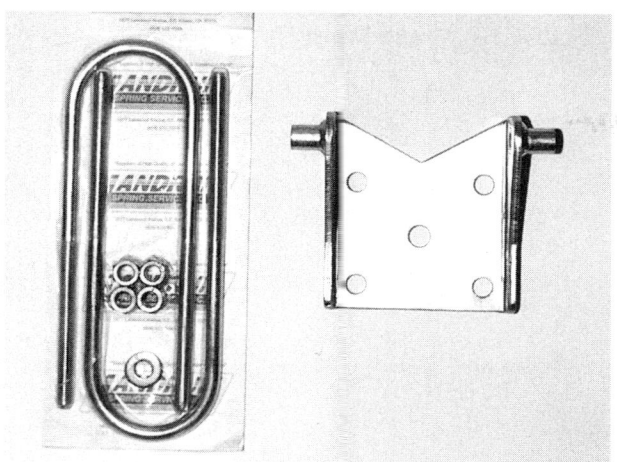

Two U-bolts and a spring plate on each side completes the installation of multileaf springs. Notice the dual-shock mounting bosses on the spring plate. This eliminates confusion about right and left plates. The U-bolts should be of good material, Grade 5 (or equivalent), and torqued to 50 or 55 lb-ft. They should be retorqued after the first race and every three to four races thereafter.

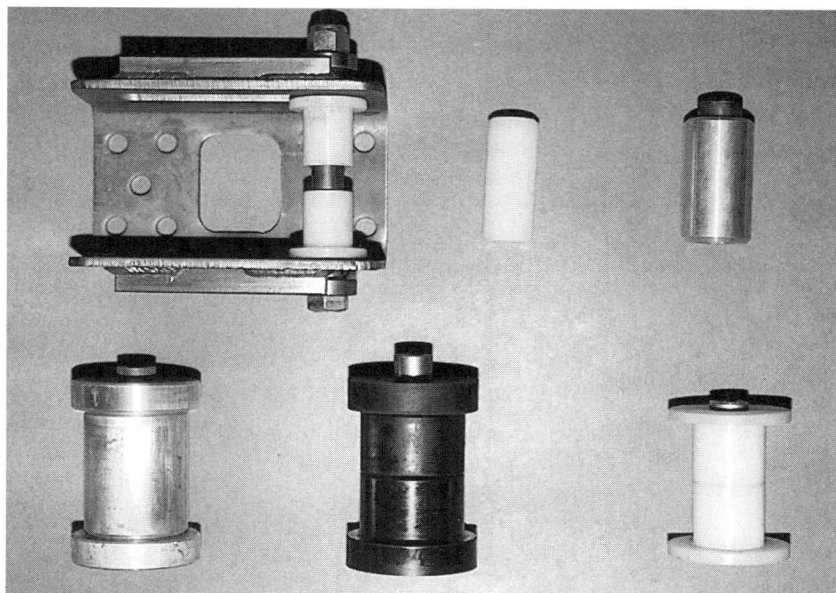

This bearing-type slider block (upper left) is shown here with its nylon spring bushings. Rear shackle bushings (upper middle) made of hard nylon or Delrin are good for reinforced shackles. Chrysler-type front bushings (upper right) made of aluminum with a rotating steel sleeve inside provide the most rigid mounting method. These must be cleaned and greased regularly. A four-piece, rigid Camaro front bushing (lower left) is the most rigid like the upper right bushing. The urethane Camaro bushing (bottom center) with steel inner sleeve makes a good replacement for stock-rubber bushing, soft enough to allow some misalignment and give, but still may be useful in rougher classes. A Chrysler front bushing (lower right) made from Delrin with steel inner sleeve requires less maintenance than aluminum/steel bushings but is still very rigid.

in-board (so that the front is narrower than the rear) adds roll-steer in the correct direction (looser during body roll when entering the turn), and tightens the chassis as the suspension rebounds coming out of the corner. Cars with low rear roll centers or high rear centers of gravity tend to stay rolled. Consequently, roll-steer induced looseness makes the car stay loose out of the corner. Here's why.

As the driver applies power, the torque reaction at the rearend housing (pinion tries to twist up) causes both front spring halves to behave like traction bars, applying downforce to the rearend and upward force on the chassis. Since the right-side spring is deflected more, its traction bar applies more down force to the right-rear tire and upward force to the chassis. Herein lies the problem. The right-rear tire at this instant is moved rearward, making the car loose. The weight transfer has increased the right-rear tire's downforce.

Now, the driver increases engine torque causing more right-rear downforce and torque to the right-rear tire patch. Remember the traction circle theory for tires? The right rear at this instant was carrying side-load, now more forward force is wanted that may exceed the limits of the tire's traction. It is very easy to cause the right-rear tire to lose its grip with the track surface, making for a very loose race car. Ideally, the additional upward force on the front of the right-rear spring will begin to "unroll" the chassis, decreasing the roll-steer and equalizing anti-

squat forces. As this occurs the car begins to tighten up, allowing more and more torque to be applied to the rear wheels. The cushioning effects of leaf springs along with smooth application of power can prevent the car from suddenly becoming loose.

What could really help this transition is something to increase the "push" in the chassis (increase left-rear weight or move the left-rear tire rearward). Of several adjustments that can be made, the easiest is increasing wedge. Crank the right-front weight-jack (wedge) bolt down a little or put a shim under the right-front spring. Increased wedge may cause difficulty in getting the car loose enough to set into the corner smoothly. Setting the car more aggressively or throwing it in sideways may be spectacular, but usually slows the car and in heavy traffic it may not be feasible. If increasing the wedge causes the car to push when power is applied in the middle of the corner, add stagger. Increasing stagger will usually make the corner entry better and help keep the car turning all the way off the corner.

Another adjustment is softening the right-rear spring one step. When the right-rear dips after the car sets going into the corner, a softer right-rear spring causes a small increase in left-rear weight, a

These mounting dimensions apply to Camaro-type springs. The height of the mounts from the ground will be determined by the spring rate chosen and the weight of the rearend of the car. The approximate deflection of the springs can be derived using the rear weight of the car and assume the springs are linear. For example, if the rear of the car weighs 1600 pounds and 200-pound springs are used, then 8 inches travel divided by two springs gives approximately 4 inches deflection of the spring at the center bolt.

stiffer left-front spring helps also. This adds wedge or "push" to the car after the body rolls, but before the driver applies the power.

Installing unequal height-lowering blocks, usually 1 to 2 inches taller on the left side than the right side can also help. When the driver applies the brakes with uneven lowering blocks, the left-rear tire moves forward farther than the right, which loosens the car. When power is applied, the springs wrap the other way, which moves the left rear backward, adding push, and counteracting looseness induced by roll-steer. A side benefit is allowing the driver to adjust braking to change where on the track and how fast the car loosens up before the body roll occurs. Remember, the roll-induced looseness (roll-steer) won't occur until the body actually rolls. This trick allows the driver to loosen the car with the brakes (or decelerating against the engine) any time he or she wants.

A subtle benefit of this system allows more static wedge to be used since the corner entry problem is reduced. The increased left-rear bite (wedge) helps keep the car straight off the corner, especially as the track gets slick and dry.

A 90/10 damper shock (cars weighing more than 2,500 pounds may need two), mounted over the top of the rearend, 5 degrees uphill (front-to-back) should be used if allowed by the rules. This increases spring life by cushioning sudden transitions from power to braking and helps keep the

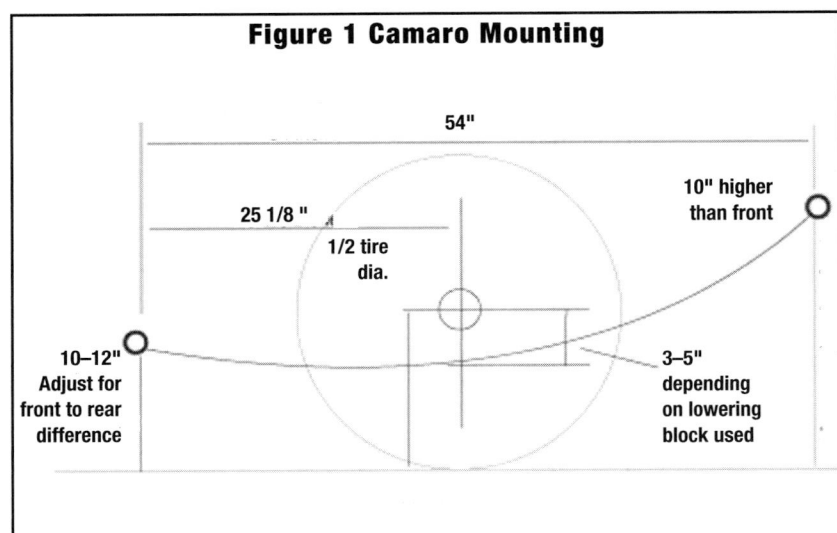

Figure 1 Camaro Mounting

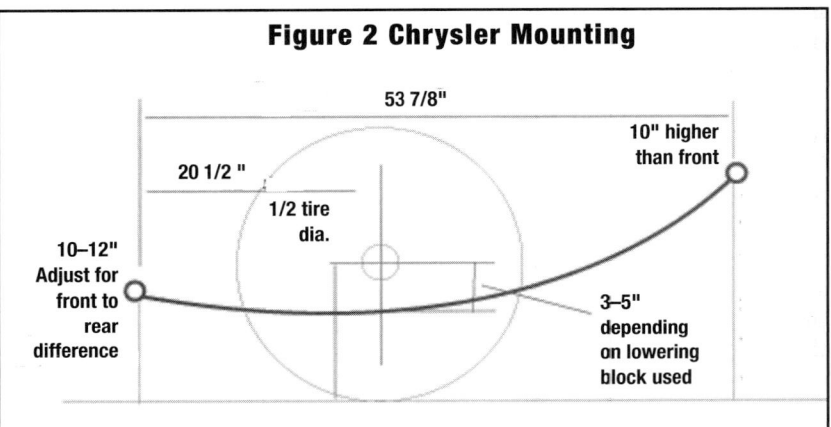

Figure 2 Chrysler Mounting

These are the Chrysler mounting dimensions. Notice how far back the rear spring eye sits. Sometimes finding a suitable mounting location is a problem for either slider block or shackles.

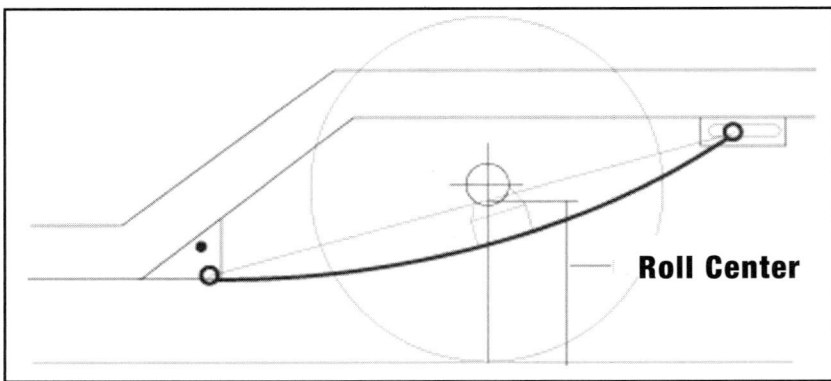

The rear roll center of a leaf spring rear is found by drawing a line between the spring eyes and finding where the vertical line through the axle center intersects. If both springs are equal in rating and deflection, the springs are symmetrical side-for-side, the roll center will be halfway between the springs. The roll center will move toward the stiffer spring if the springs are of unequal ratings.

ASSEMBLY TIP

Installing leaf springs is generally easier when the front of the springs are mounted before the rears. Be sure to lay out the mounting points according to the manufacturer's dimensions so the wheelbase comes out correctly.

With the mounts secured, set the rearend on the springs but leave the mounting pads and U-bolts loose enough to center it left to right, then put the weight of the car on the springs. Adjust the pinion angle, tighten the U-bolts, and re-check all of your settings before welding the mount pads in place.

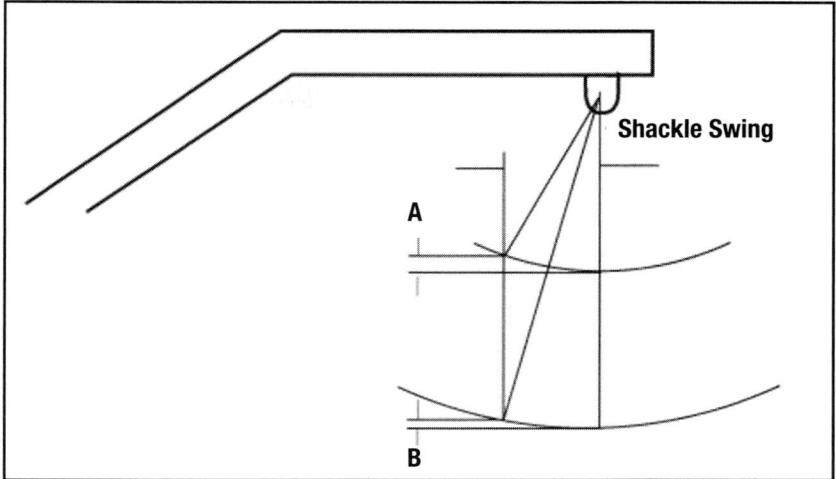

Shackle Swing

As the suspension moves up and down, a spring shackle will move through an arc causing the spring end to move up and down. This up and down movement adds or subtracts from the spring rate. Notice how on the short shackle the amount of up and down movement (A) is greater than the movement (B) of the long shackle. This weight-jacking will occur either during a bump or during body roll. Body roll weight-jacking could help or hinder depending on the direction and on-track conditions.

rearend planted going into the corner. The 90/10 shock also helps eliminate the problem of the car heading instantly toward the wall if the driver suddenly jumps off the gas in traffic.

If rules don't allow 90/10 shocks on the rearend housing, mount the left shock 3 to 4 inches forward from the axle and the right-rear shock 3 to 4 inches to the rear to help damp the rotation of the axle. Leaf springs do not have to be mounted symmetrically from right to left; moving the left spring to the left or the right spring to the left will plant the left tire harder under torque and loosen the car on braking.

Chapter 3
ROLL CENTER WISDOM

How to Measure Your Chassis for Front Roll Center Location

by Bob Bolles

Determining the roll center on your race car is critical to tuning the suspension for maximum performance.

The front roll center location is extremely important to the performance of your race car chassis. Its position will determine how your suspension will react to the forces and motions caused by negotiating a turn.

There are two reasons why we need to know the location of the roll center. One is to help determine the camber change characteristics, the other is to help determine how the dynamic forces will influence the handling of the race car. The height of the roll center (above or below the ground) affects the camber change characteristics. The position left or right of the centerline of the car will determine how the suspension will react to the dynamic forces which will cause the car to roll in the turns.

Let's measure a race car together so that we can analyze a typical double A-arm front-suspension system. We will use these measurements in a two-dimensional (height and width) geometry software program. We want to determine where the roll center is located in order to determine how that location influences the geometric and dynamic characteristics of our race car.

It's important to accurately measure these points if you expect to get accurate information back about the position of your roll center. Here are some helpful instructions on how to correctly measure the suspension points on your race car.

The first consideration is to find a level floor on which to measure. Professional teams use a surface plate made of thick machined steel. Not all of us are so fortunate. So, instead, find a portion of your garage floor that is level. We only need an area that measures about 65x20 inches to position the

front suspension over.

The measurements are much easier to determine if the engine is out of the car. If you can't get them while the engine is out of the car, try to block the car up at the chassis corners in order to provide easy access to the lower-suspension mounting points. Use the same distance to block up at each corner. Remember to support the car safely and use jackstands as backups in case the car shifts while someone is under it. You can't use too much caution here. A good distance to block up is 10 inches, because this is an easy number to subtract from your initial height measurements. It's a good idea to record the actual distance the points are from the floor if the car was at ride height, to avoid confusion later on.

We need to establish a centerline of the chassis. To correctly analyze the roll center, which influences the camber change characteristics and dynamic properties of the chassis, we want to use a centerline which is halfway between the tire contact patches. This is the centerline which the car will feel as the forces react on it in the turns. It is important for us to know where our roll center is in relation to this centerline.

Ball-joint measurements will be easier to take if the wheel is removed. Here is a method that keeps the spindle in the correct position after the wheel is removed. Before you block the car up, measure and record the length of each front shock from bolt to bolt, with the car at ride height and the race weight in it (including all fluids and the driver). Then, block the car up as explained above and remove the shocks and springs (either the spring/shock combo in a coilover configuration or the shock and separate spring in a stock-type big-spring car). In place of the shocks, insert a piece of tubing or strap metal which has two holes drilled in the ends exactly the same diameter and distance apart as the shock lengths you measured.

Mark the center on the bottom of each tire and measure between them. Place a mark on the floor halfway between these two tire centers. Repeat this procedure for the rear tires. Then, snap a chalk line over these two centerline points, front to rear, to produce the centerline you will measure the width of each point to. Now you can remove the wheels, and the spindle will be at the exact same vertical position as it was when the car was at ride height.

Now that the car is positioned so that we can get to the chassis points, we need to determine the center of rotation for each point. This is easy for Heim joints or mono ball joints, but many race cars use the OEM-type ball joints. The center of rotation is not obvious for this type of ball joint. We'll need a little help. Do not eyeball these points and guess at the location. Instead, use the information supplied by each manufacturer. Your supplier should be able to get you this information. We've included centerline diagrams on two popular ball joints. Be persistent and get this information before you start. Your roll center location will only be as accurate as your measurements.

Now, we're ready to measure. First, determine the height of each point. Measure from the floor to the center of rotation for each point. Remember to subtract the distance you blocked the chassis up from ride height from this measurement. We want a number which represents the height of each point as if the car were at ride height with all of the race weight on it.

Getting the true height of the upper chassis mounts and ball joints is going to be difficult. You won't be able to measure at right angles to the floor directly to these points. There are suspension components in the way. So, we need to be innovative. Again, do not eyeball the measurements to these points. Use a small level to extend the height of the center of rotation out and beyond the suspension components so an accurate measurement can be taken. You need a helper to get this done right.

To measure the lower control-arm chassis points, use the front chassis mounting points on the chassis. Whether you have a strut-type system or an OEM-type lower control arm, you need to use the front chassis mounting point for a two-dimensional roll center analysis.

Now that you have recorded the heights of each

Measure across the tire contact patches and divide by two to determine the centerline of the chassis, and mark this line on the floor.

Removing the shock absorbers and installing solid rods cut to the ride-height length of the shocks allows you to measure the suspension for roll center at any time.

point, measure the width each point lies from the centerline of the car. Use a plumb bob to project the lateral center of rotation down to the floor. Place a piece of masking tape on the floor to mark on. Put a crows foot mark on the tape and circle it so that it is easy to find. Measure from each point to the chalk line which marks the centerline of the race car. For this measurement, make sure the tape measure is at the right angles to the chalk line. We want the shortest distance.

Once all of the points are measured and recorded, we enter this data in the geometry software

These illustrations are taken from an engineering print of a tie-rod end and a ball joint to show where the center of rotation is on these parts. You need to measure from the center of these components when determining roll center.

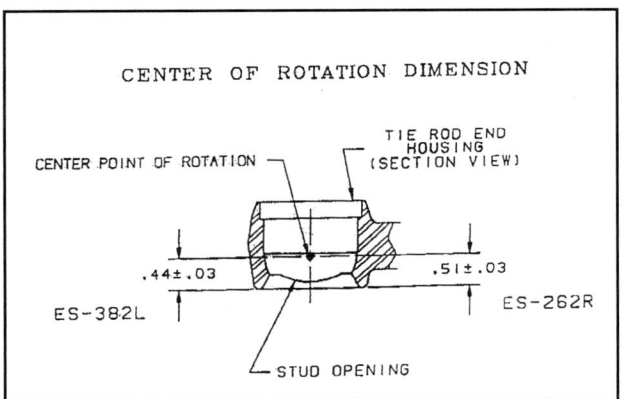

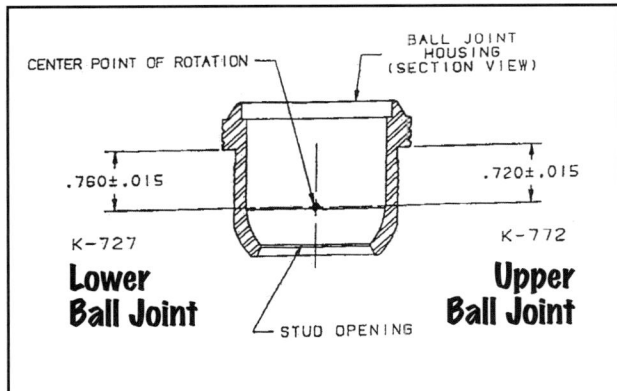

program. For this example, we use the new double A-arm geometry software produced by Chassis R&D. This Windows-based software is easy to use as it shows a picture of the cross section of the chassis. Each height and width location is entered next to the picture point on the chassis diagram. The software also allows you to quickly calculate the location of the roll centers and make changes to that location easily. The display shows the existing control-arm angles, both upper and lower, as well as the lengths of the upper control arms.

To make the changes to the roll center locations, simply type in the new control-arm angles or upper control-arm lengths and recompute the roll center locations. The program automatically calculates the new chassis mounting-point locations and tells you how far each one has moved from the original mounting points in the computer, should you decide to change them.

It's important to look at the control-arm angles which will determine the roll center location height and width (both static and dynamic). To reposition the roll centers, we must change these control-arm angles. For example, if we increase the right upper control-arm angle and/or decrease the left upper control-arm angle, we will move the roll center to the right from its original position. If we increase both the upper control-arm angles, we increase the roll center height.

The program calculates the tire camber angles after the chassis dives and rolls, and the kingpin angles of our spindles. Making changes to the upper control-arm lengths affects the amount of tire camber change when the chassis dives and rolls.

Additional pages in this software program contain provisions for calculating center of gravity, height, and wheel weights, when the user knows the percents he or she wants and the total vehicle weight.

Remember, the roll center width affects how the dynamic forces affect the front of the chassis and the height affects the amount of camber change that occurs during dive and roll. When you know where your roll center is located, you can better control how your race car will perform.

Chapter 4

DEMYSTIFYING THE FOUR LINK

Four-Link Suspension Tuning Tips for Dirt Late Models

by Donald Nosek

Only a few years ago, achieving the right setup with a four-link suspension was no easy task on a dirt Late Model, and one reserved only for the elite tuners and racers. But over the past three to four years, four-link suspension tuning has undergone some radical changes and today offers almost limitless setup possibilities for all types of race conditions. We have more ways to tune the four-link, and that's good. But that means there are now more ways to mess it up. There's certainly more to learn for a proper four-link setup, which makes it more difficult for a typical Saturday-night racer to master. However, if a race team can maximize the performance of today's four-link suspensions, they'll find themselves in the victory lane on a consistent basis.

To help explain some of the four-link tuning tips top-level race teams and chassis builders use to win races, we turned to Joe Garrison of GRT Race Cars. We also got the expert opinion of Rocket Chassis' Mark Richards, who's helped fuel the success of racer Steve Francis. Randy Jordan of Dave Poske's Performance Parts told us some tips he and driver Steve Shaver have learned over the years, and Matt Long of Clewell Racing revealed how the Ohio-based team has used four-link tuning to help them get to the victory lane in more than 248 feature wins behind veteran driver Tye Long (who has 245 of those wins) and young Cortney Clewell.

With the help of these professionals, we'll explain the different tuning components of a four-link suspension and how each change affects a dirt Late Model race car. For the purposes of this article, we'll assume most racers know the basics of the four-link suspension. Our primary focus is how to tune the suspension for specific conditions.

The four-link suspension is a complex system that the average racer has a tough time totally comprehending.

"I think where most racers miss the boat on any rear-suspension setup, especially four-link suspensions now that chassis builders place all these adjustments into it, is they overadjust and make the car tighter," says Mark Richards. "That happens 90 percent of the time for most local racers. They are under this big misconception that they can drive the push out of the car, but you can't because these cars simply have way too much traction available."

Traction is important in dirt Late Model racing, and that's one of the primary functions of the four-link suspension, but the answer to every tuning problem is not always more traction. If the race car is too tight, the driver may actually have to break traction just to get the car to turn. That results in no traction or control through the turn, and the driver is then forced to find some trick to regain traction to exit the turn. All this serves to slow down the car rather than speed it up.

Here you can see a great view of the left-rear suspension setup on this Rocket Chassis dirt Late Model. The top link is mounted at an upward angle in the center-hole position on the chassis, while the bottom link has no angle but is shorter in length than the top link. This setup enhances side bite but also controls roll-steer through the corner.

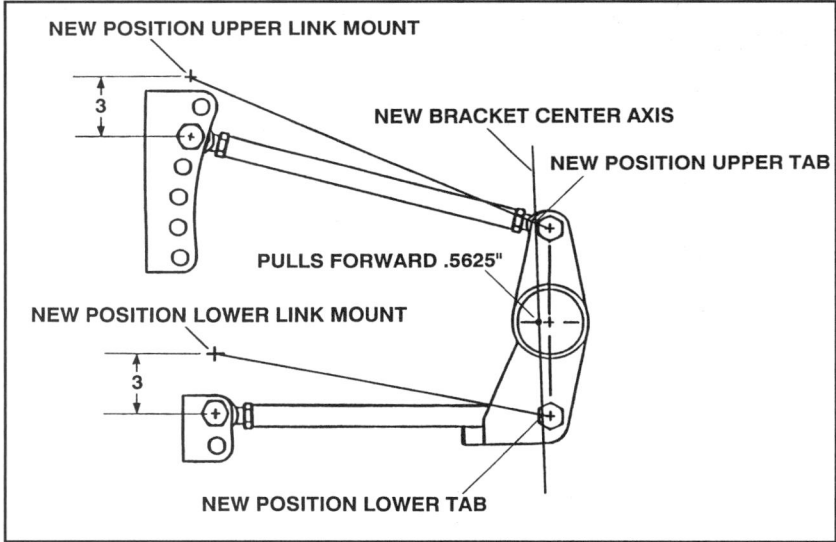

This illustration shows how adjusting the four-link bars can actually change the wheelbase of the race car. For instance, if a racer moves both links up three inches, the axle will move forward about a half-inch. When tuning a four-link, it is essential to understand what each adjustment means in the overall picture.

"The key word is momentum," says Richards. "When you have momentum, you don't need nearly as much traction."

Racers run into the same problem of lost momentum when they attempt to give the car too much induced roll in hopes of gaining more bite.

"I have found that on this four-bar stuff, Friday- and Saturday-night racers have a real bad conception of what a four-bar should do to a race

car," says Randy Jordan of Dave Poske's. "They think it's traction. They think there's magic in them bars to give them traction. And there's not." Instead of looking for "magic," Jordan suggests racers find a neutral setup that works, then start moving the four-link suspension's tunable components one hole at a time until they understand what works and what doesn't on the race car. He also suggests not trying to tune too much on race night, or racers might end up chasing their tails and missing the setup entirely.

"With some of these guys, the racetrack will be heavy when they arrive. Their car is pushing because they had it set up for slick conditions last week. They'll want to come in and change bars and everything," he explains. "Well, the track's going to be back the way it was last week, or close to it, later in the evening. If they know they had the bars the way they liked last week, they should be tuning the wheel offsets, stagger, air pressure, and stuff like that—and leave the bars alone!"

FOUR-LINK BAR ANGLES

It's the heart of the four-link suspension, so we might as well start with the four links themselves. Racers have a variety of ways to tune the bars, but there are basically two main items being adjusted: the rod angle and the rod length. And within those adjustments, the upper bars are usually concerned with corner exit, while the bottom bars mostly affect the entry.

"The more upward angle you have, top or bottom, the more weight is going to be induced through the trailing rods to the rear axle, making more traction," says Richards. "But at the same time, you have a factor of roll-steer that has to be considered. Too much upward angle will cause too much roll-steer."

During a roll-steer condition caused by top links that are sharply angled, the right-rear tire actually moves rearward and the left-rear tire forward, causing the rear axle to steer toward the outside of the track. If this roll-steer is too excessive, handling the car will be too difficult and actually negate any advantage gained by increasing the bite. However, some roll-steer can actually help the race car through the turn. The trick when tuning the angles of the upper bars on a four-link suspension is finding a happy medium between traction and loose roll-steer on different track conditions.

"On the bigger tracks," says Richards, "where you have to carry more momentum in the turns, less upward angle is needed." Conversely, for a short heavy track, where racers need to get lots of bite to exit the turn quickly, the upper links need to be angled more sharply—between 15 and 20 degrees.

"The angles we try to run on the upper bars are anywhere from 15 to 18 degrees for a normal track, but if it gets slick we can set the angles anywhere from 20 to 24 degrees," says Long.

As stated, the steep angle also creates more loose roll-steer, but there is another concern with severely angled bars: out-tuning the shocks. "The higher you adjust the left-side upper link, as the car goes into the corner and rolls over, it lets the rear end roll forward and shove up on the car," says Jordan. "If that happens, the shock and spring aren't doing a whole lot, and you're using mechanical rods to apply traction to your race car. You have to be careful with both links on the left side. Don't get them at such an angle that it tops the shock out. When it tops the shock out, it'll actually jerk the left rear tire up off the ground."

To control entry into a turn, racers will look to tune the bottom links instead of the top. "Let's say you're loose going into a turn on the right, you can lower the bottom angle and shorten the bottom bar length," says Long. On a slick, short track, limiting the roll-steer by angling the lower bar either level or negative up to five degrees will help the race car get through the corners faster.

But even finding the right balance between traction and loose roll-steer over a certain track is sometimes not enough. Racers also have to consider high and low lines around the course and factor how each plays into the four-link setup. Imagine you've found a nice groove around the track, the setup is perfect, and you're flying high on the outside—but you run into traffic. To pass, you might have to forsake your high line and dive to the bottom. If you have the suspension set up only to run wide, you might not be able to drive under a car and pass coming out of the turn. That's why it's best when tuning the four-link suspension to have a setup that is adaptable for the driver. Often, that means a setup that is more neutral so the driver has a better chance of manipulating the race car through different lines around the racetrack.

FOUR-LINK BAR LENGTHS

In conjunction with altering the bar angles, racers have the option of tuning the length of the bars as well. Some racers we talked to never change the length of their bars once they find their ideal setup; they choose only to adjust the angles and other factors from week to week. On the other hand, some of our experts admit they may change the length of the bars at the track to adjust for conditions.

In general, if the right-bottom bar is shortened and/or the left-bottom bar lengthened, the car will

This photo illustrates the increasingly popular left-rear shock-behind setup (shock and spring), with a stabilizing rebound shock (shock without spring) mounted in the front as well.

Notice the straight Panhard bar attaching to the left side of the pinion. This is used in instances where the racetrack is slick, with big, slow corners.

tighten up on entry—similar to the effect lowering the right-bottom bar angle has. In contrast, if the right-bottom bar is lengthened, the car will become looser on entry.

"Our normal bottom-right bar is around 14 to 15 inches, and we might go as short as nine or 10 inches to tighten the car up," says Long.

"If we get to a track that is wet and heavy for qualifying a speed race, and our hot laps are real tight," he continues, "we'll just throw our standard long bar on the bottom right, which is about 15 inches long. Then if the racetrack slicks up for the feature, we'll put a shorter bar back on the bottom right."

As for the top bars, those lengths are usually left between 17.5 and 18.5 inches but can be different lengths on the right versus the left. Joe Garrison of GRT Race Cars actually runs longer bars on the left and slightly shorter ones on the right to prevent the car from becoming too tight. He says this "bar lead" should be anywhere from a quarter to a half inch difference in the length of the bars.

"On a track with sweeping corners, you probably want just a little more lead in the bars, versus a track that is real tight and stop-and-go," says Garrison. "Trailing rods that are too equal in length can cause too excessive roll-steer."

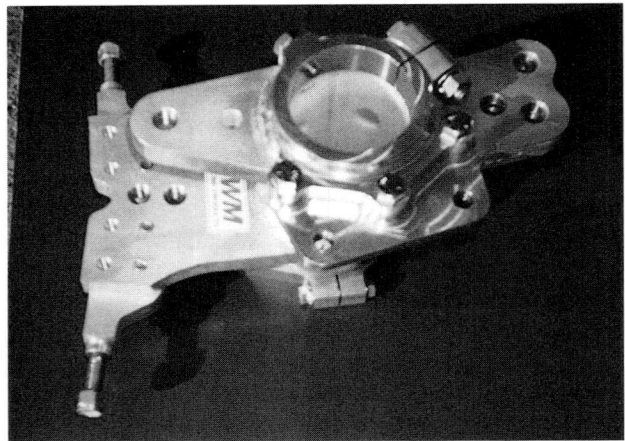

Here, instead of a straight Panhard bar, Rocket Chassis has used a J-bar that wraps around the rearend and attaches to the right side. Notice the number of setup holes available, allowing racers to tune by changing the angle of the J-bar.

What you'll find on most dirt Late Model setups these days is a long upper link and a shorter bottom link. This setup induces traction but helps limit roll-steer. Figuring Saturday-night racers already have enough to worry about with tuning the bar lengths, this may be a feature most choose to setup once then never touch again.

SHOCK AND SPRING SETUP

When you talk shock and spring setup on a dirt Late Model, the possibility for tuning with double-adjustable race shocks and a variety of spring rates is a monumental task, and one that probably deserves an entire book unto itself. But there is a recent development in the position of the shock that warrants attention, enough so that Mark Richards credits it with "saving the four-link race car." We're talking about moving the left-rear shock behind the axle. Developed about several years ago, this setup actually places the shock and spring behind the axle, instead of its traditional location in the front.

In general, the left-rear shock has a similar effect increasing traction as another tuning setup racers have used for a long time: clamped-up. "If there is traction available on the racetrack surface, the left-rear clamped setup provides lots of forward traction," says Richards. "But when there's no traction available, the left-rear shock-behind setup is body-roll controlled, where the left-rear clamp is throttle controlled. So you don't have to have quite as much traction on the race surface for the shock behind to work. All you have to do is cause the car to have some sort of roll. That's why you see a lot of the cars today rolled onto the right side through the corners, with the left side almost airborne."

What racers mean by clamped up is instead of attaching the shocks in the standard position on the birdcages, leaving them floating with the axle, the shocks are clamped straight to a fixed position on

One of hottest items in dirt Late Model racing is an aluminum birdcage. Unlike the beefy cast floaters of the past, these billet pieces come with a variety of adjustable mounting options and save nearly 11 pounds of unsprung weight over their predecessors. This example from TWM Racing Products is made from 6061-T6 aluminum and can be ordered with numerous options, including floated or clamped brakes, various coilover mounts, and more.

the axle.

As Long describes it, clamped up is "instant traction." "Or you can run double-clamped," says Richards, "in which both shocks and springs are mounted to the axle. That allows both rear wheels to be loaded when the torque is applied to the rearend. Now one of the downfalls to that setup is there has to be some traction there for the tires to make adhesion for that setup to work."

Essentially, forcing more traction into the rear tires will only work when there is traction to be had. Otherwise, running double-clamped on a slick track might leave racers spinning their wheels, literally.

"We find on a double-clamped setup that it works best in extremely heavy, short-track conditions where you need the right-rear wheel to drive equally with the left rear," says Richards. But with a large number of racetracks, the high cost of digging them up these days seems have influenced operators to leave their tracks slick, even at the beginning of the night. Therefore, a double-clamped setup is rarer than in years past.

"We still run left-rear clamp setups quite regularly. And where we find that a left-rear side clamp setup works well is on a tacky, usually short racetrack," Richards continues. "The reason why is because a clamped-bracket setup is controlled by the throttle. So it makes a real drivable setup on a short, tacky racetrack. We use the unclamped setup mostly on high-speed momentum racetracks, where we don't need quite as much bite."

"If it gets really slick, the problem with clamped

up is there's a lot of roll-steer," says Long. "When you hit the throttle, it pushes the left rear clear out from the car. That's why many of those guys have just switched to the left-rear shock-behind setup."

Although a left side clamped is still effective and widely used, a left-rear shock-behind setup puts a lot of diagonal weight force in the car, which allows the driver to steer off the left rear and right front. It also helps increase traction, which is why this setup has become so popular. But a racer can't just decide to throw the shock behind the axle and expect the car to be faster—there's more to it than that.

"With your shock behind, the biggest problem you have is usually being too tight coming off the corner, but you also have a lot of bite," says Long. "That can be a dangerous combination. To combat that, racers can adjust the spring rate on the left rear, taking wedge out of the car."

Richards says, "With the shock and spring behind, a high-wedge setup seems to work well most of the time on a high-speed racetrack. On a high-wedge setup, the springs need to have more split between them. Let's say you ran a 300-psi left rear and a 200 right—that might be a good split for a high-wedge setup. However, on a low-wedge setup, the springs need to be more equal, like 250 and 225 psi."

You can also adjust the shocks to help stiffen the right or left rear—and it's all quite simple if you have a double-adjustable racing shock. With those advanced shocks, you can easily tune rebound and com-pression between heat races or practice laps.

"What happens is when you put the shock behind," continues Long, "you want maximum torque but also the least resistance so that the spring pushes the shock out. Everybody started off running what's called a dummy shock on the front of the axle."

While the shock behind has a spring around it, the shock in front serves as the control shock, limiting the distance that the rearend can fall down, and therefore does not need a spring. "Basically, what we use that shock in front for is to control suspension travel," says Richards. "We can help control the body roll, how much the car raises or drops, with the left-rear shock in front on a shock-behind setup. Sometimes you'll even see guys use a zero, dead shock on the backside and use the shock in front as the control shock. The control shock in front actually moves at a little bit greater speed than when it's used with the spring."

PANHARD BAR OR J-BAR

In a dirt Late Model race car, this bar controls the car's roll center. While it has the same function, its name depends on its shape, which depends on where the bar is mounted to the pinion. A short, straight bar mounted to the left side of the driveshaft is a Panhard bar. A J-bar extends over to the right side of the pinion and is bent down in a "J" shape to mount lower on the bracket and maintain the desired angle depending on track conditions. Basically, all J-bars also are Panhard bars, but not all Panhard bars are J-bars. However, for dirt Late Models, J-bars are used in just about every situation except a slick, big, slow corner racetrack.

"By raising the J-bar up on the frame to create more of an angle, it makes it easier for the car to roll over onto the right side," says Long. "If you're drifting across the track going into the turn, then raise the chassis mount for the J-bar, and it will help the weight transfer to the right side."

"We use a 19 to 20 inch, and it has probably anywhere from seven to nine inches of angle in it for our standard setup," says Garrison. "By angle, I mean the mounting point on the pinion versus the mounting point on the chassis. If the track is extremely slick, you need a lot of angle in the J-bar to help induce roll in the car and create side bite and drive. On a bigger track where you can carry more speed, you don't need as much angle in the J-bar. You might even use a longer bar, say as much as 25 inches. On a real tight, stop-and-go racetrack, we might even use a short, straight Panhard bar if we have to get the car to turn quickly."

Garrison continues, "Actually, on our standard J-bar setup, with it mounted on the right side of the pinion, I think it helps tighten the car on entry but allows it to turn through the middle and coming off. In contrast, the straight bar that mounts on the left side of the pinion makes the race car more loose on entry and gives the drive coming off—perfect for a short, stop-and-go racetrack."

Again, you can see that while the basic principles behind the suspension are rather simple, a lot of variables are involved when it comes to tuning the four-link. Each adjustment on one end affects several factors on the other side. Unless you understand what how each change affects the rest of the race car, you might end up counteracting a change you just made.

If there's one theme most of these experts tried to espouse in our interviews, it was to make sure you understand it's best to take things slowly when tuning the four-link. Remember, it's momentum—not traction—that wins races.

ROLL-STEER VS. BITE

Like most race car tuning, where almost all changes come with countereffects, an adjustment to a four-link suspension to increase traction likely will have an adverse effect on roll-steer. In contrast, correcting roll-steer usually means losing some bite. And on top of that, the same tuning change on one side usually has an opposite effect on the other. This sidebar helps explain why these two oppose each other in a four-link suspension.

Note: In the illustrations below, the dotted lines represent the link paths on the right side during compression, while the solid lines represent the left side during rebound.

While a certain amount of roll-steer may be desirable to induce side bite, too much can make the race car difficult to control in the middle and exit of the turns. Notice the incredible amount of lift the left side of the car has in this picture. If the race car rolls too far over on the right side, it can lose the most important force in racing—momentum—and take longer to exit the turn.

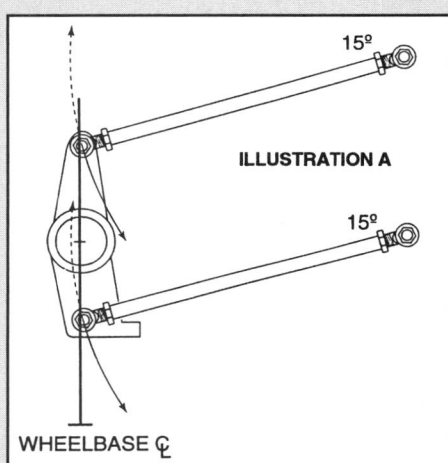

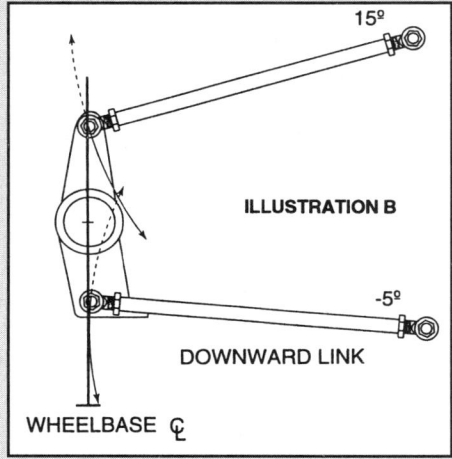

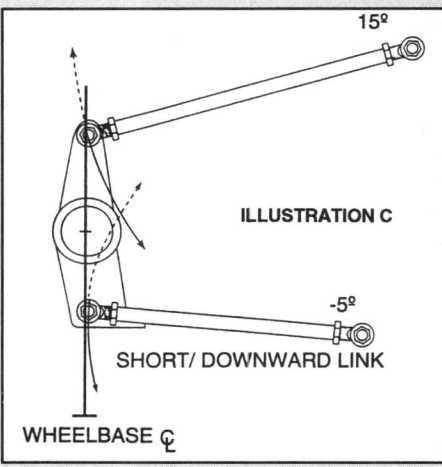

While a standard street car would likely have the upper and lower link bars mounted parallel, at zero degrees, dirt Late MS2 model racers can angle the link bars upward to increase side bite. In Illustration A, the links are equal and mounted at an upward angle on the chassis. During cornering in this setup, the body roll will force the right-side links to push the axle rearward, while the left-side links will pull the axle forward. The result is improved side bite but increased roll-steer.

In Illustration B, a dirt Late Model racer might leave the top link-bar angle upward but lower the bottom link angle to zero degrees or even a few degrees negative. This adjustment produces axle thrust to load the rear tires from the top link. At the same time, lowering the angle of the bottom link reduces the rearward movement at the bottom of the birdcage and reduces roll-steer in the chassis.

One of the best ways to tune roll-steer is to change the length of the links. This drawing is similar to Illustration B, but the length of the bottom link is shortened. With a shorter bottom link on the right side, the arc at the birdcage becomes sharper, which pulls the left side of the axle forward even more and further reduces roll-steer.

Chapter 5

FOUR-LINK DIRT TECHNOLOGY

Design Goals for Today's Four-Link Rear Suspension

by Bob Bolles

A four-link dirt Late Model rear suspension is designed to have a large range of rear steer in either direction. This full range of adjustability allows the racer the opportunity to make adjustments for the changing conditions that occur on dirt surfaces. The attitude of the car on dry, slick tracks can be quite radical, and we will study why this might work.

The parallel four-link rear suspension is common to dirt Late Model and dirt Modified cars. This suspension is sometimes found on asphalt cars, but not too often. The original purpose for using the four-link suspension was to end up with a suspension system that produced very little rear steer as the chassis moved vertically. Racers, being true to their nature, decided to experiment with various angles in the four-link and found that zero rear steer did not fit all circumstances.

Today, we have various schools of thought on where to position the links on a four-bar system and many more theories on why to do it. Let's examine what happens with the various changes and look at the big picture to try to understand what is really happening to our cars. Even though the current trend among top dirt Late Model racers is to minimize the steering characteristics of the rear suspension, there are times when these teams must get radical.

We will be analyzing two basic designs of the four-link rear suspension: the "standard" four-link (that we will call a four-link) where both links are forward of the rear end axle tube, and the "Z" link where the top link is rearward, and the bottom link is forward. Both of these designs can be positioned to produce near-zero rear steer. The forward design can be made to produce somewhat more rear steer than the Z-link.

Both of these suspension types are usually attached to a birdcage that is not locked to the rear axle tube, where the rear end is free to rotate. A separate structure is attached to the rear end to control rotational movement upon acceleration and braking. This could be a "third link" similar to that used on a three-link suspension, a lift arm that runs forward and is attached well in front of the rear end or a combination of several systems. If the link brackets were mounted solid to the rear axle tube, then as the car rolled in the turns, there would be a significant amount of binding going on. The suspension would be trying to twist the rear end as each axle tube would be rotated in opposite directions.

If we change the angles of the links so that one produces more movement at its end of the birdcage, we can move that end in a direction that will cause the rear end to steer away from straight ahead. This is called rear steer, and in most cases it is not desirable, especially on asphalt stock cars. Rear steer on dirt is not only more acceptable, but downright necessary under certain conditions.

To even begin to understand how the car will rear steer and to what extent, we first need to completely understand the movement of the chassis and what causes this movement. The chassis mounting points of the four-link

If the right side of our suspension travels up or down in the turns, we will see near-zero rear steer by positioning the links as shown (reverse the vertical arrows and the horizontal arrows will be reversed too). The same would be true for the left side. The vertical movement of the chassis will cause the top and bottom of the birdcage to move in different horizontal directions, resulting in minimal movement of the rear end.

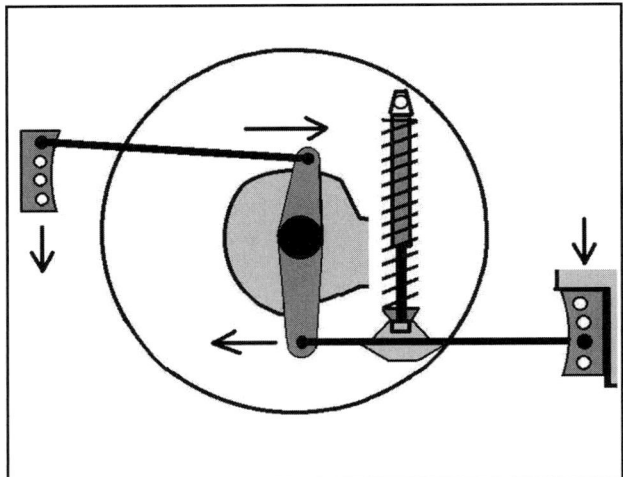

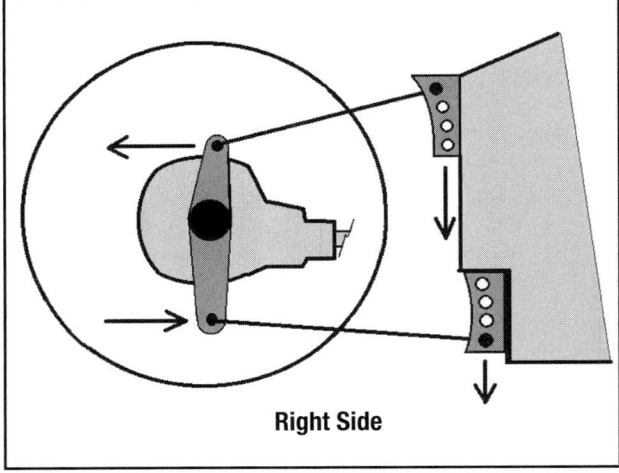

Right Side

will move vertically as the car transitions into and out of the turns and even down the straightaway; the degree of movement dictates the degree of rear steer.

As a chassis rolls in the turns, three basic things can be happening overall: 1) the left side of the chassis may move up, and the right side may move down, 2) the right side may move down, and the left side may stay near that static location, and 3) the left side may move up, and the right side may remain unchanged. With the same set of suspension link locations at each side, a car may well produce very different rear steer characteristics from each of

the three scenarios.

A four-link can be made to produce varying amounts of fore and aft movement of its end of the rear axle in either direction, depending on the combined angles of the links. If we start at a neutral setting for the links, meaning that for a certain range of movement up or down, the axle will not move fore and aft, let's see how we can produce axle movement.

If we move the chassis mount for the bottom link up, then the rear axle will move forward as the chassis moves up. On the Z-link, we see the same effect for the bottom link. The opposite is true if the chassis moves down. For both systems, the axle would move to the rear. That is exactly why we need to know which way the chassis is moving at each side of the car under all conditions.

Knowledge of the extent and direction of shock travel will come in handy as we plan out our rear geometry. We can translate shock movement to suspension movement. Shock travel indicators or data acquisition will tell us what is really happening.

A direct influence on chassis movement in the turns is the J-bar or Panhard bar. If the bar is mounted more parallel to the ground, then it will have little influence on the vertical location of the chassis in the turns, and the chassis will move similar to examples seen earlier. If there is a lot of angle in the bar with the left end higher than the right end (chassis mount to the left side as is most popular), then as the car turns left, the bar angle will have a jacking effect, causing the left side of the bar to want to ride up over the right side of the bar. This movement would raise the entire rear of the car.

Here we see Shannon Babb's car raised up on the left side, but there appears to be little rear steer, as the left-rear tire is still centered in the wheelwell. This would indicate that the links are positioned for minimal rear steer on this four-link car. Note in the blowup that the tire appears to be biting very hard. Another indication of less rear steer is the more straight-ahead attitude of the overall body.

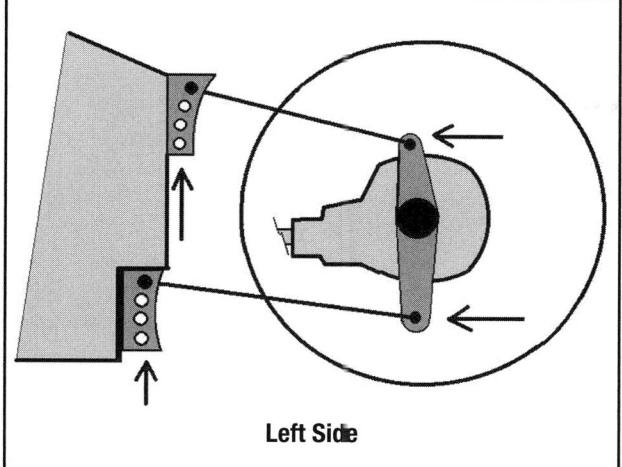

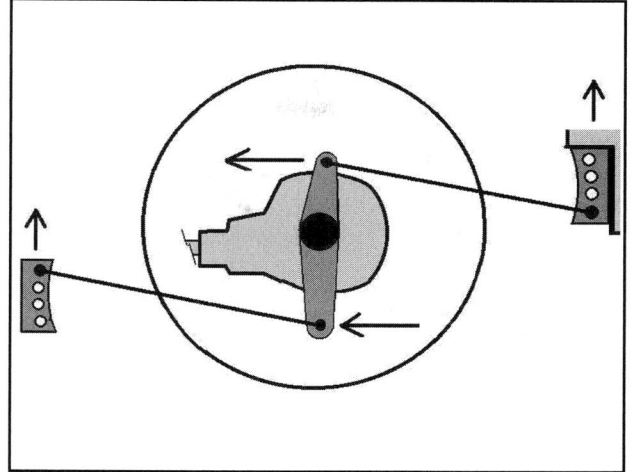

Putting the chassis mounts for the links in these positions will cause maximum rear steer as the chassis moves up in the turns. We can dial in various amounts of rear steer by moving the connections to different holes for each or both links.

Under those conditions, if the car rolls (and we know it does), and the whole chassis rises up, then the right side links may well remain in their static locations producing near-zero rear steer at that side. On the other side of the chassis, there will be a combined vertical movement of those links to where the lift associated with the bar angle will be combined with the roll lift so that they can produce a large amount of rear steer at the left side of the axle. This movement usually pulls that end of the axle forward, and the rear end will steer to the right.

Upon acceleration off the corners, the rear end will be driven forward, and if the forward ends of the links are higher than the rear ends, there is a further movement of the chassis to a higher level.

If we just look at the way the car steers, we might conclude that this is not a very good idea. On

asphalt, this would produce a very loose car that would be undriveable. But if we look at the whole picture, including the aerodynamics of the body, we start to see why our lap times may be lower by doing this on dirt, especially on a very dry, slick racetrack.

Winged Sprint Cars will generally run at an angle to the direction they are moving through the turns. The tall sides on the wing catch a lot of air and will produce a lateral force that is the opposite of the centrifugal force that tries to take the car to the wall. To make a long story short, the aero force counteracts the lateral g-force and helps the car go faster through the turns, just like having more tire grip.

The combined effects that raise the whole rear of the car also put the rear spoiler higher into the

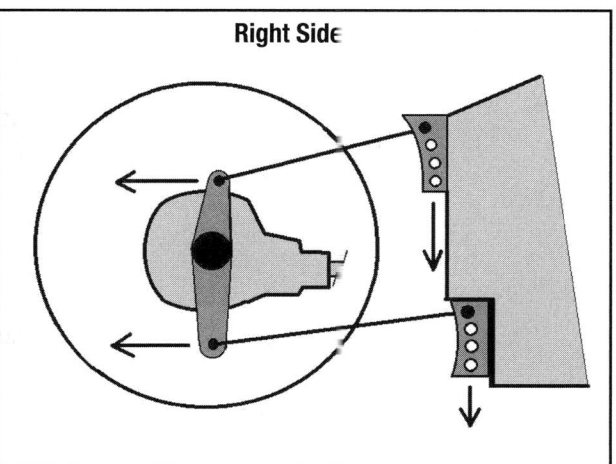

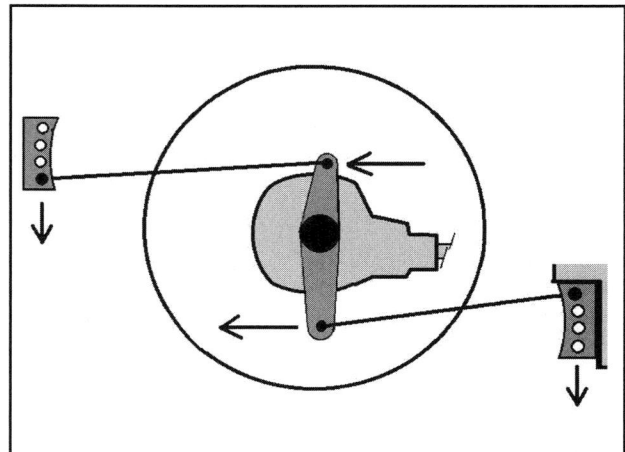

Should the right side of the chassis move down in the turns, positioning the links as shown will create rear steer to the right. We will need to know if the right side of our car is moving down or the left side is moving up, or if there is a combination of both. When we know that, we can properly plan our rear steer design.

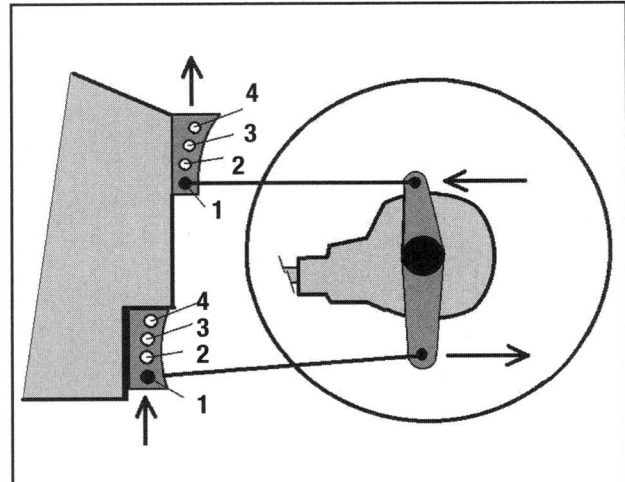

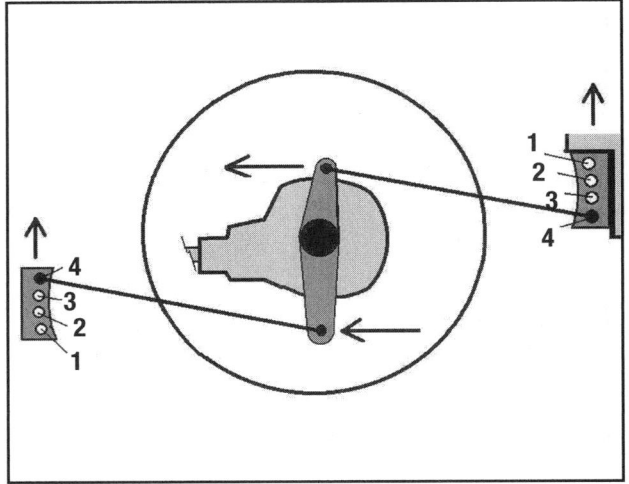

If, for each suspension system, the number "1" position produced minimal rear steer or movement of the axle at that side (which it should), then if we move from numbers 1 through 4, we would see more rear steer as the left side of the chassis moved up. The four-link left side view shows how moving the links up through the numbers produces more axle movement by causing the birdcage to move forward at both ends, top and bottom. The Z-link shown is a little different because the top link chassis mount must move down to create the same effect. This shows how going up in number moves the axle to the front, helping produce more rear steer by causing the birdcage to move forward.

windstream; that can produce more aero downforce at the rear. This helps give us more traction to provide better bite off the corners.

If the track has a lot of grip already, we need much less rear steer and the associated aero help. So, we make changes to our rear links so that less rear steer occurs. We may even benefit from creating opposite rear steer, to the left, to gain more rear traction, just like we do on asphalt. The operative word here is change. We must be willing to make changes, and a more thorough understanding of what happens with each change will make it easier to do with better results.

What happens to many dirt tracks is they get wet and tacky as the day starts out. A car that is jacked up and rear steered to the max just won't get through the mud as well as one that is more level with all four tires on the ground. We can position the links so that there is very little rear steer for these conditions.

As the track dries out and becomes more slick, we may need more rear steer and rear jacking to get the rear spoiler up into the airstream for more rear downforce. The rear steer has more effect on the angle of the body related to the direction that the

car is traveling, and an aero side force helps pull the car to the left to keep it from sliding. Putting more angle in the J-bar is warranted now.

Once we fully understand how link angle changes produce rear steer at each side, we can make helpful adjustments at the track as the conditions change. We need to plan which changes to make and be able to do them quickly with little effort. A setup sheet that shows which holes to mount the links for each set of conditions would help the crew make fast changes. If you don't want your crew to know why you are doing things (secrecy is an asset at times), then just number the different sheets and tell them to set the car to sheet "3"—period.

Because we need to adjust other parameters on the car for changing conditions, we can include spring rate changes, shock changes, and fifth and sixth coil changes as well when we make up the setup sheets. Once we develop our setup sheets, as we race the car, we can tweak the numbers according to the results. The process of dialing in the car to the conditions using our setup sheets may take a few races, but at least we will have a plan.

Chapter 6
BASIC SPRINT CAR CHASSIS TECH

by Bob Bolles

Setting up a Sprint Car is all about the process. There are slight differences between the winged and wingless cars, but much is similar. We will explore some new ideas about setup.

Sprint Car setup information is very hard to come by. We searched Google under "Sprint Car setup," and "Sprint Car technology," and what we got was a reference to Sprint Car schools, books and videos. Those will definitely help you get started, but what about the current trends in setup? Doesn't our approach to setting up race cars ever change? Sure it does, even with Sprint Cars.

Even on the 4M.net website, usually a treasure trove of good information shared by weekend racers, we did not see much discussion about specific setup trends. We saw some suggestions made to contact the manufacturer for specific spring/torsion bar setups, blocking heights, and so on.

We did find an interesting reference by the Midwest Sprint Car Association (www.msasprints.com) with the following statement: ". . . with the club rules focusing on driver ability and car setup, the MSA has been an attractive option for drivers looking for fun racing at a reasonable cost." It seems as though the Sprint Car sanctioning bodies encourage experimentation with car setups, so let's experiment.

We always like to think more on the fringe of common thinking and explore areas of setup that may not be either popular or common. This is the way racing has evolved ever since the very beginning, and we certainly want to continue and encourage that long-standing trend. Let's take a look at

the different areas associated with Sprint Car design, construction, and setup.

BASIC DESIGN OF A SPRINT CAR

All Sprint Cars are constructed in a similar fashion. The suspension is straight axle for both the front and rear. The driver sits in the center of the car. The rear axle is supported inside birdcages. The rear tires are much larger and wider than the fronts, especially the right rear, and the cars are sprung with either a coilover spring/shock combination, torsion bars, or a combination of both.

The steering is basic straight-axle steering that was used in the early 20th century passenger cars where the steering box is located just in front of the steering wheel and connected to the front axle by a link that goes from the steering arm attached to the box up to a steering arm that is attached to the spindle. A tie rod connects both front spindles. The cars may have wings or not, depending on the particular rules. This all sounds elementary to a veteran sprinter, and you may say, "So what? We know all of that." I am pointing out these various design features so we can evaluate them individually later on.

The two basic problems Sprint Car teams have associated with handling are the inability of the car to efficiently turn the corner, especially on asphalt or tacky dirt, and a lack of

Sprint cars have large and wide right-rear tires, and the resulting large amount of stagger helps make the car turn and provides more bite when exiting the corners. There are some other ways to help the car turn better that we will explore.

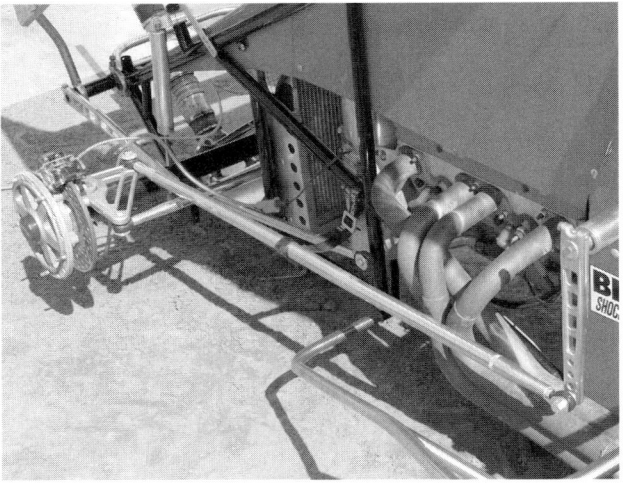

The steering system for most Sprint Cars use a steering box located just in front of the steering wheel, a connecting link that runs from the box steering arm up to the front spindle steering arm. As the car rolls, the movement of the rear of the connecting link causes the front wheels to steer.

forward bite coming off the corners, especially on dirt when it is dry. These cars usually have a large power-to-weight ratio, meaning a lot of available power and not much weight. The wide right-rear (RR) tire helps provide forward bite but can also make the car tight on entry and in the middle of the turn. If we loosen the car to make it turn, we lose forward bite. It seems to be a vicious circle.

If we use the standard manufacturer's suggested settings for setup, we might not find the true potential for our cars. All car builders must publish information that would fit a wide range of tracks and track conditions. If your team runs a particular track every weekend, you should work to tailor your setup to fit your racetrack.

Let's evaluate and propose ideas for the different components associated with Sprint Cars and maybe "attack" some of the normal thinking. You do not have to agree to or follow as gospel on what we present, but you do need to do your own thinking about your car and make reasonable changes that might make your car handle better and be faster.

SPRINT CAR STEERING SYSTEM

There is a big problem with the steering systems on a typical Sprint Car. As the car dives and rolls in the turns, the steering connecting rod pushes or pulls on the steering arm at the spindle causing the wheels to steer, even when there is no steering input by the driver.

A team once showed me a picture of their wingless Sprint Car, racing at Indianapolis (Indiana) Raceway Park, taken at mid-turn, and the front and rear wheels were running straight ahead, but the driver had the steering wheel turned to the right about 45 degrees. This may be an extreme example, but maybe not.

In this case, the radius rod may have been set at level or with the rear end lower than the front end. As the car rolled to the right in the turns, the rear of the link moved down, the front end of the radius rod, as well as the steering arm at the spindle, moved rearward, and the wheels would turn left. The driver then had to compensate by introducing steering to the right in order to go straight ahead. He might have thought the car was sliding when in reality it was tracking very neutral.

This is a hard problem to solve with the current design. Someone needs to put a lot of thought into redesigning the system. If we had our way, the ideal system would not steer at all during body roll. One idea would be to mount a rack-and-pinion onto the straight axle that was activated with a steering shaft with a sliding fixture to accommodate the motion of the straight axle.

The rear axle is located fore and aft by both a forward control arm and also a connection to the torsion bar arm at the birdcage. This Z-link arrangement, if properly designed, can help keep the rear from rear steering during dive and roll. Note the Watts link assembly behind the axle and shock and the birdcage outside the control arms.

Here, we can see the connection of the end of the torsion bar arm to the bottom of the birdcage at the end of the axle tube. The torsion bar provides both spring rate and fore/aft location control of the axle.

The front straight axle is located fore and aft by parallel radius rods. This helps to maintain caster in the axle while the car is rolling and moving vertically. Note that the torsion bar arm is resting on top of the axle tube. The shocks limit the amount the axle can rebound.

Another design might incorporate a flexible cable system that would not be influenced by the motion of the chassis. This system would be much like the ones used in boats. It could even be of a hydraulic design. The "sending" unit would be at the steering wheel and the "slave" unit would be attached to the front axle and be able to move the drag link directly or, in a more traditional way, move a steering arm at one of the spindles. This would eliminate roll steer and make the car much more stable, especially when racing on asphalt.

REAR STEER IN A SPRINT CAR

The control arms, or trailing arms, are somewhat short in a typical Sprint Car. That, combined with a relatively narrow track width, means there could be some amount of rear steer as the chassis rolls and moves vertically. If parallel forward-mounted arms are used, there will be an arc created when the axle moves. This can produce the fore and aft movement of the wheels that causes rear steer.

On some torsion bar systems, the torsion bar arm is bolted to the birdcage and acts like a trailing link to form a Z-link. These arms are indeed very short and considerable movement of the birdcage occurs as the car rolls and bumps or dives, but the rear axle does not move fore and aft to create rear steer.

If the torsion arms were mounted to a Heim connector and the rear end was controlled by forward-mounted trailing arms, then much less rear steer would occur. The angle of the trailing arms is important too. As we set our ride heights, we need

to be careful as to the angles of the trailing arms.

SPRINT CAR SPRING RATES

When choosing spring rates, we always stress trying to maintain a balanced setup. The Sprint Car has a different set of suspension systems than a stock car. Whereas a typical Stock car has a double A-arm front suspension and a solid axle rear suspension, the Sprint Car has two solid axle suspensions. This means that each can be sprung with similar stiffness if the mounting points are similar and the weights are similar.

In reality, there are small differences in the spring mounting points as well as the heights of the Panhard bars or other lateral location devices. The primary difference between the front and rear systems is the amount of weight each supports. This is important in planning how we should spring our cars.

In a typical Sprint Car, the front-to-rear percent of total weight is 35–40 percent front and 60–65

The different lines on the track provide more or less grip. The lower line is darker and has more moisture, providing more grip. The upper groove is lighter in color and is dryer; that means it has less grip. Either way, we need to develop a balanced setup to help keep the left-front tire on the track to help turn the car.

This adjuster for ride height on a torsion bar system is located at the opposite end from the arm. Here we see the adjuster for the left-rear chassis height. Changes to the ride height also are changes to the weight distribution.

percent rear. With more weight supported by the rear suspension, we would imagine the spring stiffness would be greater at the rear. Traditionally, all Sprint Car teams install stiffer springs at the front than the rear. It almost sounds like things are being done backward.

This reverse "theory," as such, has been evaluated and some Sprint Cars have been set up, whereas the rear springs were stiffer than the front springs, or at the very least the same stiffness at both ends. These cars had very good results.

Remember early on we stated that these cars were typically tight and hard to turn. If the rear is sprung too lightly for the load it carries, excess weight transfer will occur at the front, and the left-front

tire will have much less weight on it at mid-turn, or in a worse case scenario, be lifted off the track surface. A three-wheel car will not turn as well as a four-wheel car.

I helped to design a pushrod Sprint Car where the two suspension systems had pushrods, similar to the open-wheel formula cars, and the springs and shocks were mounted inboard. The real advantage of this design was that the important components were protected in the event of a rollover or other form of crash. The pushrods were simple and easy to make and replaced in minutes if needed.

This car was sprung stiffer in the rear. For whatever reason, it won its very first and only race it ran. It was banned from competition from then on.

On these Florida Mini Sprints, one uses a coilover front suspension and the other uses torsion bars for the springs. Each can be adjusted to provide a dynamic balance between the front and rear suspensions.

The point here is that it can work. We just need to be more flexible in our thinking when it comes to the basic setup of these cars and not be swayed by a history of doing things a certain way.

WEIGHING THE SPRINT CAR

Most Sprint Car teams do not scale their cars. Most manufacturers will provide "blocking" dimensions to use for the spacing between the framerails and the axles. These are set distances to be used with suggested spring rates so that the wheel weights will be right for the intended use.

Some teams will scale the car to use as a baseline, especially when the car runs well. They will try to repeat those wheel weights in the future. No matter which way you set up the weighing of your car, certain processes must be used to ensure consistency and accuracy.

I witnessed a team member blocking his car at the racetrack on a dirt surface that was very much uneven. There is no telling what the actual wheel weights turned out to be from this crude process. There is a better way to do this.

The car should be blocked or weighed at the shop on a flat concrete surface. Ideally, you would find four spots that are the same elevation to roll the four tires onto. Scales can be leveled by using aluminum plates as shims. The most critical leveling is across the front and across the rear scales. Front-to-rear slope, if kept to a minimum, will not make a noticeable difference in scaling.

It is a good idea to either scale or block the car before and after the races. You can keep track of the handling versus wheel weights that way. We sometimes happen across something that is better and unexpected. If we leave the shop with our usual 48 percent of cross, and through a series of changes at the track end up with 51 percent, should the car get better, we might have made the discovery that we need more crossweight in the car.

SPRINT CAR HANDLING BALANCE

The handling balance for a Sprint Car needs to be a dynamic balance just as in a Stock car. All four tires need to be in contact with the racing surface

Large wings with large rectangular side panels help produce a lot of downforce as well as lateral force for the winged Sprint Car. We often see two wings: one smaller at the front, and the big one over the driver with its added load more centered on the wheelbase.

and have the most load ending up on them as possible. If we have a truly balanced setup, the left-front (LF) tire will carry a decent load and the car will turn better.

For a winged Sprint Car, it would make sense that our spring rates, side to side, would be the same. Spring split on a solid axle suspension has a dramatic effect on roll stiffness. Since the cars roll first to the right on entry and then left at mid-turn, we need to have the same roll stiffness in both directions. If a stiffer RR spring over the left-rear (LR) spring promotes roll stiffness in a roll to the right, then the reverse would be true for a roll to the left.

Our handling can be much different from corner entry to mid-turn if we run different rate springs on each side for winged Sprint Cars. The Panhard bar height, or Watts link moment center height, must also be tuned to the spring stiffness as well as the spring base. The idea is to develop a setup for your Sprint Car that has the same roll characteristics for chassis roll to the right or left so that the handling balance is equal all of the way through the turns.

The wing on a Sprint Car can be made adjustable by hinging the supports and adding a hydraulic cylinder. The driver controls the angle of the wing by moving a lever near his left hand. Note the lever just to the left of the steering wheel, above the knee brace, and at the end of the braided hose coming from the cylinder (arrow).

WINGED VS. WINGLESS

The setups for the winged Sprint Cars versus what is referred to as wingless is necessarily much different due to the high amount of downforce produced by the wings. The spring stiffness must be more for the winged cars, mostly on the cars racing on asphalt. The speeds are much greater when Sprint Cars are racing on asphalt and with wings attached. The loads that the tires will experience are much higher than the wingless cars.

More loading means more traction, and the g-forces go up considerably with the increased speed through the turns. This necessitates a higher overall spring stiffness and in some cases, a spring split with the right-side springs being more rate than the left sides when running on asphalt.

Here too, we see a definite need for chassis setup balance. We really want all four tires to carry maximum load. At the LF, we need the tire to carry load to help the car turn. Dirt cars can go sideways to point the car off the turns; asphalt cars don't have that luxury.

With Sprint Cars, as with all circle-track race cars, do not be afraid to experiment. Use the manufacturer's baseline to start with and then branch out and try different setup parameters. When something works, keep it. When something does not work, note it and do not try it again. Racing technology has evolved since the early days, and we don't want to stop that process now. Everything that will make you faster has not been discovered. That's one of the things that make racing so much fun.

DIRT CHASSIS TUNING

Chapter 7
WEIGHT MANAGEMENT

Place It Right for Bite

by Scott Bloomquist
Photos and Illustrations by Tom Hintz

Scott Bloomquist has won several Hav-A-Tampa Dirt Racing Series championships. In this chapter, he shares his knowledge of weight management to help you win.

Weight may be one of the most talked about subjects in racing; often focusing on ridding the car of a few extra pounds. Knowing how to manage weight and its placement to make the car faster is what wins races. Some racers get their cars so heavy during construction that adding weight to balance the chassis puts them way over the minimum regulated by their rules. Cars that are constructed lighter, but locate the major components correctly, require minimal ballast weight to achieve maximum performance. Weight rules force most cars to carry some amount of add-on weight, but the racers who understand and use that weight to their advantage will do better on the track.

Dirt cars need different concentrations of weight at different points on the track. Going into the turn, left-side weight should transfer to the right-side tires to produce side-bite, which makes the car take a set (stop sliding). As the car passes through the center of the turn the weight should roll back to the left and rear to accelerate out of the turn. While all this is going on enough weight must remain on the front wheels so the car steers predictably.

COMPONENTS

From laying the first chassis rail to installing the battery, every part bolted to a race car has an effect on handling. Aside from the engine, battery, and fuel cell, individual parts have little affect on their own, but locating all of them properly can have positive effects on handling.

Dirt racers generally mount the major components quite a bit higher and more centered than asphalt racers to help

promote side-bite. Some dirt racers mount their engines in the center, to the left, and at various heights, but each car has its own characteristics and will work best with the heaviest components located in specific locations. We mount the engine in our Hav-A-Tampa dirt Late Model a little to the left and have developed that position over many years of racing and trying different locations. Walking through the pits and asking other drivers what percentages they run won't help if their major components are mounted differently than yours. Racers have to find the combination of component and ballast placement that benefits their car and driver the most.

The first step is following the chassis builder's recommendations because they developed specific component locations to work with their design. Our team runs Barry Wright race cars, and over the years we have spent a lot of time working to find the best placement for the components in all the chassis's he sells. If we mounted identical engines and fuel cells in our Barry Wright chassis in the positions used by another chassis builder our car wouldn't work nearly as well.

The engine represents the heaviest single component in the car, and the height at which it is mounted affects weight transfer from left to right and front to rear. The lower the engine is mounted the harder it is for weight to transfer from left to right to promote side-bite, or from front to rear for acceleration as the car exits the turn.

Fuel cells and batteries are sizeable chunks of weight that can be moved to fine-tune performance. Dirt racers often mount fuel cells in the center of the chassis, but some move

GENERAL DIRT WEIGHT PERCENTAGES

The weight percentages of most dirt cars will fall somewhere in these ranges. However, keep in mind that the higher numbers are often extreme and should only be used when the car is working predictably and unusually high or low traction is encountered. Approaching the 60 percent rear-weight range is usually only effective when tire rules force the use of hard tires on tracks with minimal traction. Getting the car to turn with this much rear weight can be very difficult. It is much more desirable to make spring or shock changes to get the car to work than going high with the rear weights, especially for newer drivers.

Left: 51–55%
Right: 45–49%
Front: 40–51%
Rear: 49–60%

Generally speaking, when required by rules to run front weight near the engine, mount it as high as possible, unless racing at a high-traction track.

them slightly to the left. In many cases, when moving these heavy components to the left they also have to be mounted higher to produce the same amount of side-bite.

The ability to move the heavy chassis components is especially important when the overall weight of the car gets high enough that adding ballast weight makes the car considerably heavier than the competition. Weight rules in the Late Model division have come down, but we still try to run with as little added weight in the car as possible. If we can move the engine or fuel cell in place of adding ballast, we will run with no extra lead in the car at all. To help accomplish this our cars are built so the fuel cell can be raised or lowered 4 inches. Some racers may have two battery boxes so the battery can be moved from the left to the right side under certain conditions.

WEIGHT & SPEED

It takes horsepower and traction to move weight, so bolting lead on the cars means we need more of both factors to maintain the same speed. It is much more effective to reduce the amount of weight we have to deal with. Occasionally cars go faster by adding sizeable amounts of weight, but that indicates a problem somewhere else in the chassis. Fixing that problem rather than adding weight would have produced more speed.

We sometimes add weight before a 100-lap race to replace weight lost by the amount of fuel burned off, but this depends on how we expect the track to

change throughout the race. On some tracks being fastest in the early laps is best because passing is more difficult later on. Other tracks require the car to handle best at the end of the event to win. Either way, racers have to determine how much of a change keeps them working their best without giving up too much at another point of the race.

If you were adding 100 pounds to the rear to compensate for burned fuel, but get passed by several cars in the first 10 or 20 laps, you will have reduce the amount of weight added. Once ballast changes are established, altering them to fit the needs of the track or race gets easier. Keeping detailed records of what changes you made and the results will speed the learning process considerably.

The ideal situation is building the car so when it is complete with the driver in place, the weight percentages are optimum without additional ballast weight. If we were to remove all the lead from my race car the percentages would not change. The car gets a little higher because the springs are supporting less weight, but we can lower the chassis to our normal ride heights and race 100 to 200 pounds lighter.

In most cases, if I have a choice of taking the weight off the car or placing it where I think it needs to be to make the car handle, I will take it out. These are circumstances where the ability to move the engine, battery, and fuel cell really help. If the weight percentages can be altered to suit the conditions by moving existing components we can

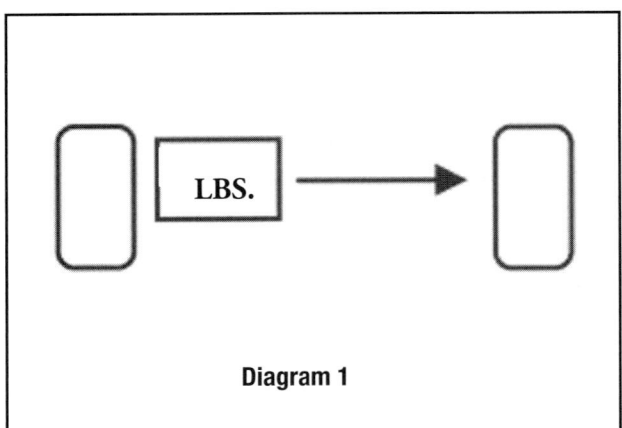

Diagram 1

Weight located very low tends to move parllel to the track rather than transferring to the right side loosening the car on corner entry.

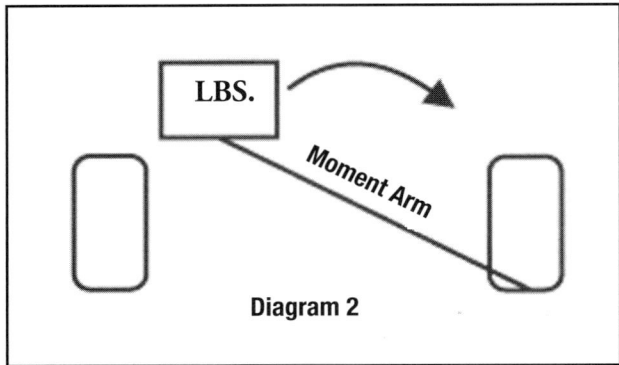

Diagram 2

Raising the weight increases the leverage exerted by the moment arm, increasing the amount of weight transfer and resulting side bite.

eliminate adding weight to the car. Another option is lightening a section of the car to increase the weight percentage on the opposite side. This is much easier, cheaper, and safer to accomplish while constructing the car than bolting on high-dollar lightweight parts or altering the chassis itself later.

DRIVER PLACEMENT

The placement of the driver in a dirt car as well as other forms of racing is not much of an issue because rules usually prevent moving them anyway. We have to look at driver placement as a constant and concentrate on his or her actual weight and how that impacts weight percentages and transfer. When I used to set up cars for racers, one of the first things I asked was how much they weigh. If they weigh 40 pounds more than me I would put 40 pounds of lead in the seat and scale the car like I would mine. Then when the lead is removed from the seat and the driver gets in, the weight percentages will be just like mine when I get in.

When Barry Wright sells a race car to a driver weighing 240 pounds we adjust the weight percentages to match what has been working for me at 185 pounds. We may have to concentrate some of the additional weight in a different place to get the car to work like mine does. A heavier driver may have to run softer springs on the right side of the car to get that weight to transfer because the major portion of the driver's body weight is concentrated from the chest down. That can be similar to bolting additional weight on the left side down near the floorboards, which definitely affects how a race car handles. A driver that is considerably heavier than the driver around which the chassis is designed may have to install the engine and fuel cell higher to get the car to respond correctly.

LEFT-SIDE WEIGHT

Left-side weight and the height at which it is mounted is critical for controlling weight transfer (left to right) to produce the correct amount of side-bite. Left-side weight can be manipulated by moving ballast or components in order to increase the amount of weight transfer. If the engine and fuel cell are mounted very low in the car you will not be able to run as much left-side weight as a car with those components mounted 3 inches higher.

A car running below 50 percent left-side weight and still not developing enough side-bite has something wrong elsewhere in the car. If we ran 50 percent left-side weight in our car it would turn over, but each car will respond to a different left-side percentage, usually in the 50–55 percent range.

WEIGHT AND HEIGHT

Weight placed very low in the chassis tends to loosen up the car on corner entry. Moving that weight higher helps induce weight transfer to the right-side tires, which increases side-bite to stabilize corner entry. Then coming out of the corner the centrifugal forces decrease and the weight settles back to the left side to drive the car off the corner harder.

Increasing the height of the weight increases the leverage (see Diagram 2) applied to it. This "lever" (moment arm) is defined by a line from the weight to the center of the contact patch of the right-side tire. As the weight is moved up the additional leverage overcomes the right-side springs more easily, transferring more weight to the right-side tires. Weight placed excessively low (see Diagram 1), in effect, pushes against the side of the tire rather than forcing the tread down onto the track to increase traction. When weight is located too low it no longer rolls onto the right-side springs, but rather

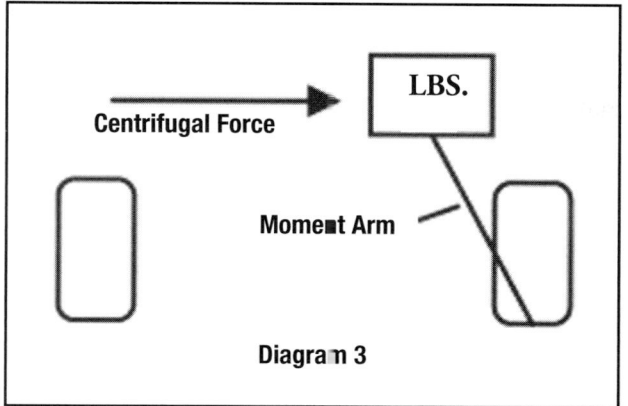

As weight is moved to the right the leverage of the moment arm increases quickly. Weight transfer and side-bite can be increased without adding additional weight.

Late Model racers are notoriously picky about fellow racers getting a peak of their rear suspension and weight placement.

tries to move the chassis parallel to the track surface.

The key is using the weight moment arm to get enough weight transfer from the weight already in the car, rather than adding more weight. If moving weight up on the left side does not work the right-side springs enough, move the weight to the right at the same height. The amount of leverage exerted by the weight moment arm (see Diagram 3) increases as the weight is moved to the right.

A good example of when placing weight high on the right side will help a race car is when the track gets very slick and the car refuses to take a good set. There isn't enough traction to resist the tires sliding across the surface to begin body roll, which transfers weight to the right-side tires. Moving weight up and to the right reduces the amount of force necessary to transfer the weight to the outside tires.

Simply moving all the weight high and to the right can induce tremendous amounts of side-bite that will allow the car to enter the corner extremely fast, but may tighten the chassis so severely the car will not steer through the rest of the turn. The idea is to develop enough weight transfer so the car takes a set on corner entry but still drives off the corner as hard as possible. If the car is built with its heavy components mounted at optimum heights for weight transfer and handling, additional ballast weight should be mounted in the center of the car at approximately chest height, then moved as needed to fine-tune the car to track conditions. Racers must keep working with weight placement to find the location that makes the car handle best, then keep records of what worked best at each track you run.

TUNING WITH WEIGHT

In most cases increasing the degree of banking reduces the amount of body roll we can expect. As body roll decreases so does weight transfer to the

right-side tires because centrifugal force pushes the weight down on all four tires more than transferring to the right side. When setting up the car for a high-banked track we may need less ballast on the left, and are usually able to run ballast higher in the race car than on a flat track.

Moving weight should be used as a fine-tuning tool to get the car working as best it can for the track conditions. If our car is a little loose going into the turns we may raise all the weight 6 or 8 inches. Then if the car is still loose on entry we start moving the weight, at the new height, to the right. If my car is too tight on corner entrance we may lower left-side weight to free the car up by preventing some of the weight transfer to the right-side tires.

Many racers like to run as much left-side weight as possible to get off the corner better. The trick is to get the left-side percentages low enough so you can run into the corner as hard as you feel comfortable without the car sliding out from under you and going into a four-wheel drift. Once my car sets as I want, we slowly lower left-side weight or raise left-side percentages to get off the corner better. Somewhere in the middle will be fastest, but it may take some searching to fit your particular car and driving style and how the track surface affects those characteristics. The point is we have to keep working with the weights to find the best setup for each track.

FRONT TO REAR WEIGHTS

Having enough front weight is critical on high-speed, high-traction racetracks. When there is a lot of traction available we may take some of the weight off the rear tires and put it on the fronts to

There are various weight placement and mounting methods. Follow your track rules about weight mounting and protect you and your fellow racers from stray weights on the track.

gain steering. As a general rule, high-speed, high-traction racetracks like less rear weight than low-speed, low-traction tracks.

Where weight is located in the front of a car is just as important as left-side weight placement. Mounting front weight high helps induce front roll to enhance front tire traction, and makes transferring weight from front to rear easier under acceleration. Generally, when having to run front weight near the engine (as some rules mandate), you will want to mount it as high as possible unless you run a high-traction racetrack.

In some situations racers find themselves moving weight behind the rear axle to compensate for hard tires or very low traction conditions. While this can work, it is a double-edged sword. As weight is placed farther behind the rear axle, front-wheel weight decreases and steering control is reduced. Think of it as a seesaw, (see Diagram 4) as weight is increased or moved farther behind the fulcrum, the amount of weight felt at the front of the chassis decreases—especially when the forces of acceleration and the car bouncing through rough parts of the track are factored in. Locating the weight directly over the rear axle will produce forward bite while helping to maintain steering control through the corner.

Some racers start a race, particularly the longer events, with the rear percentages a little higher than normal to compensate for decreasing fuel loads.

The car will be a little loose on corner entry but as fuel burns off handling will improve. Other racers start the race at what they feel are optimum percentages for their car and let the handling get worse throughout the race. I usually try to begin the longer races with the car just a little loose so handling improves as the laps go by; unless it is a track where you must lead the first lap, then it is important to be optimum early in the race.

TRACK CONDITIONS & WEIGHT

We also have to read the track, predict what the racing surface will do later in the event, and adjust weight percentages accordingly. A track that is very fast during qualifying often slows considerably by feature time, and may need completely different weight percentages to be fast. Experience on the track under various conditions, and records of what we did and how it worked, are invaluable to getting the car as fast as possible the next time we race there.

Some racers, particularly those that have their car scaled by someone else, go to the track with a specified set of chassis changes to be made between qualifying, heat, and feature races. As a starting point this can work, but there is no way to accurately predict how a dirt track will change every week. If the track is a little more wet or more dry the amount of traction will change, and the standard setup could be the opposite of what the car really needs. Learning to manipulate weight

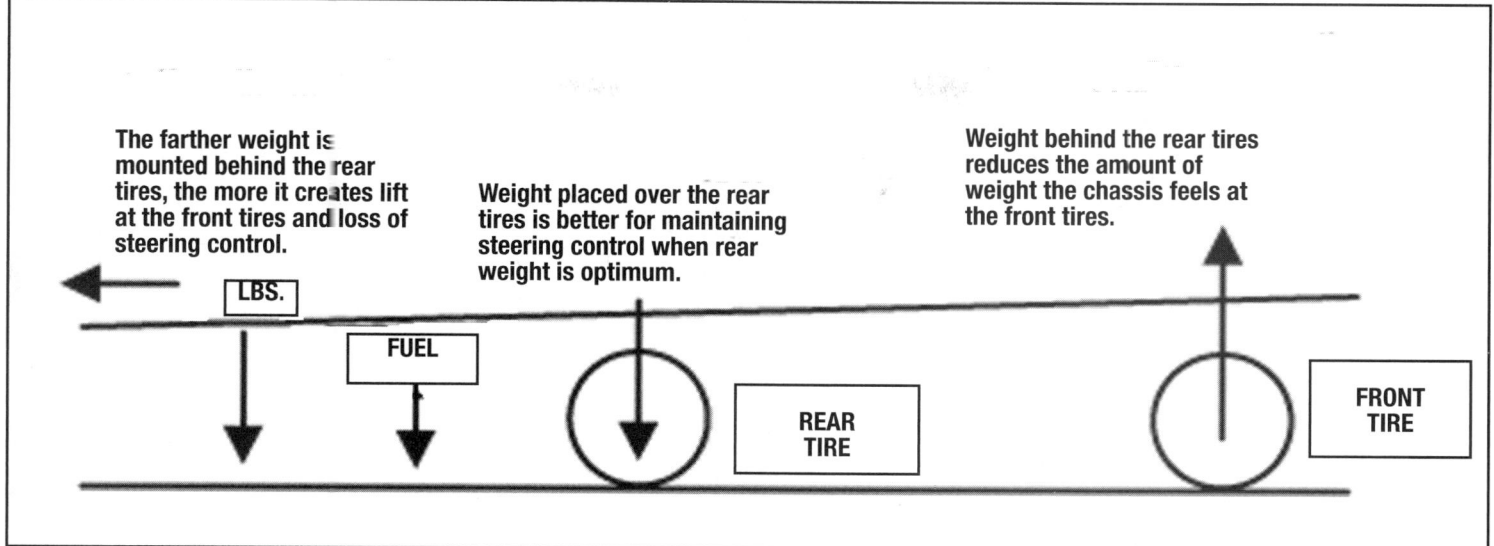

The farther weight is mounted behind the rear tires, the more it creates lift at the front tires and loss of steering control.

Weight placed over the rear tires is better for maintaining steering control when rear weight is optimum.

Weight behind the rear tires reduces the amount of weight the chassis feels at the front tires.

LBS.

FUEL

REAR TIRE

FRONT TIRE

according to the various track conditions is a sure way to consistently increase your performance.

Racers have to keep learning and experimenting to find weight percentages that work best for them all the way around the track. Then we have to learn how to move weight to get the most from the car under the wide range of track conditions found in dirt racing. Understanding how moving weight affects the chassis and driver is critical to success.

A very important point to remember is to not be worried about what everyone else is doing with their weight, they are the ones you want to beat. Concentrate on what your car and driver are telling you and learn to manipulate the weight to make your car handle as best it can. Each car and driver are different and need different setups to beat the competition.

Chapter 8
WORKING WITH SHOCKS

by Billy Moyer and Tom Hintz
Photos by Tom Hintz

Billy Moyer's ability to read a track and predict how it will change through the night, combined with his knowledge of shocks, helps to make him one of the most consistent drivers to ever strap on a dirt Late Model.

Until recently, choosing shocks for a race car was pretty easy. There were a very limited number of different shocks and, for most tracks, there were a number of them that were commonly run on the front, and another for the rear. The ability to tune the car for different tracks or conditions was yet to come. Then, as racers learned more about how shocks could help their cars, the manufacturers began producing more valving options, but their number remained very small. But that is changing fast in dirt racing, especially over the last year or so. Over the last few years, Winston Cup teams have generated a lot of press by using their shocks for chassis tuning; but we have to remember that it was the short-track racers who first began experimenting with using shocks to improve lap times through handling.

For a long time, nearly everyone in dirt racing used 7600 shocks on the front and 9500 shocks on the rear. The biggest difference between them was the length. A 7600 shock had a 7-inch stroke, and a 9500 had a 9-inch stroke, but both had 50/50 valving, or, equal resistance in compression and rebound. This confined our chassis tuning to changing spring rates, geometry, and tire pressures.

Years ago, the number of valving variations were even more limited and some teams discovered that using more than one shock at one or more wheels helped get the car around the corners faster. Today we have a much wider range of valving options from which to choose, and before long, fully adjustable and maybe even rebuildable shocks, similar to

what they have in NASCAR, will come to short-track racing. Over the last year or two there has been a lot of work done with shocks on dirt that eventually will give dirt racers a much wider range of shock options with which to further tune their chassis and make their race cars handle better than ever before.

COMPRESSION VS. REBOUND

The shock industry has really expanded its efforts to produce split-valve shocks—shocks with different valving for compression and rebound. Installing shocks that react to a different amount of force during compression then rebound at each corner of the car has opened the door for more precise tuning of the chassis over a wider range of situations. Weight transfer can be timed and better controlled with specific compression or rebound valving to help get the chassis doing what we need it to do and when. Whether you use coilovers or factory-style shocks, advancements in valving are going to make a big difference in your racing.

I run Bilstein shocks that are gas charged, which greatly reduces foaming of the shock oil. Reducing foaming reduces the shock fade we used to get later in the race and makes the chassis much more consistent. This consistency means we can make chassis adjustments and not have to guess how much the shocks will change over the course of the race.

As shock manufacturers increase valving combinations, racers are better able to choose shocks that more closely meet

Notice the difference in the attitudes of these two cars. The car in the lead has considerably more body roll established early in the turn to help get the most traction possible on the dry-slick track. Using less rebound resistance on the left-side shocks helps encourage this body roll.

When the track is wet and tacky, many teams like holding the left side of the car down by increasing rebound stiffness. This can help get more drive from the left-rear tire. Remember that the major adjustments are made with the springs and the shocks used to fine-tune them.

their particular needs. The Bilstein shocks I use currently come in 14 different valvings in compression and 14 in rebound. Such a large number of variations allows us to make very small shock changes that complement the rest of our chassis setup.

SHOCKS AND SPRINGS

Originally, shocks were used primarily to stabilize the wheels and tires by damping the springs in order to smooth and control their reaction when the tire encountered irregularities in the racing surface. Without the shocks' damping properties, a spring would cycle through compression and rebound oscillations long after the bump was encountered. In racing, this would mean the chassis would constantly be moving up and down, loading and unloading the tires unpredictably, and never allowing the car to settle or to take a set in the corners.

We have to remember that spring rates control the ride heights of the chassis, and how much the chassis raises or lowers during cornering or going over bumps. By controlling the extent of chassis movement, the springs also control how much weight is transferred. Shock valving allows us to control how fast the chassis goes through that range of movement.

TUNING WITH SHOCKS

An important point to remember is that not all chassis are designed to be adjusted alike. Some

chassis builders use a lot more wedge in their cars than others, which means that changing the right-front spring or shock could have very different results on their chassis than on another design. Before making wholesale changes to your shocks, find out from your chassis builder what general changes they recommend to cure basic handling problems such as one side pushing or one running loose. Then you can fine-tune those adjustments with shock changes.

Because springs control so much of chassis tuning, shocks will always be a fine adjustment. If your springs are wrong for the conditions, shocks will not, on their own, remedy the situation. Put the right springs in the car and proper tuning with shocks can make you faster than the competition.

As with all chassis changes, we need to go slowly and keep records of how the chassis reacted. Very often, small changes in the compression or rebound rate of a single shock can have dramatic results on how the car gets around the track. There will also be times when changes will be hard to notice. If your car is already handling pretty well, small changes in the shocks should not put you out of the ballpark altogether if you go the wrong way on valving. If your driver is consistent, making such a mistake gives you more information about what the car likes, and helps to define the direction you need to go with valving in the future. Recording this information for later use is the only way to accurately predict what you need to do with shocks for all the different conditions we can run into on a dirt track.

This is a portion of the shock inventory Billy Moyer carries in his hauler, in part because of the long distances and extended time he spends on the road. Though he generally doesn't often change his shock setup drastically, racing against the best in the business means he has to have the right parts available.

These are the shocks Moyer's crew laid out for a night's racing. Keeping accurate records and lots of experience helps narrow the number of shocks they need in the attempt to get the car hooked up.

DRIVER FEEL

A mistake many teams make is trying to duplicate the setup which the driver of another car may like. Each driver has a little different style that needs a different setup. But more important is the feel each driver likes and is most comfortable with on the track. If you can get the chassis set up to where the driver feels very comfortable, the performance is almost certain to increase.

I like to be able to feel the car roll up on the right-front tire as I enter the corner, then feel the car rock back onto the right-rear tire, sort of like a teeter-totter. When we can get the car like that, I always go fast. But before you try setting your car up, remember this is an individual preference and it took me a long time to develop that feeling and to find out how to set our car up to get it. Along the way there were some really scary rides and finishes below what the car was capable of. Your driver may

like a very different feeling to go just as fast. Being able to fine-tune the shocks is a great help to finding the right feel for your driver.

SHOCKS AND TIMING

One of the things that complicates tuning with the new kinds of shocks is the ability to have different valving on compression than on rebound. This configuration provides a much wider range of adjustments but also demands that we be right in our choices in both directions of travel.

By varying the amount of force (valving) necessary to compress or extend (rebound) the shock, we can change the rate at which the chassis raises or sinks as the car goes around the track. If we increase right-side compression stiffness and decrease left-side rebound stiffness, the weight will transfer to the right side faster. By reversing the stiffness values of the shocks, we can slow the weight transfer. The rate of the springs still regulates how far the chassis will roll, but the valving of the shocks helps control how fast the chassis goes through that range of motion.

Using shock valving to control the rate of chassis movement allows us to time where, in the corner, maximum roll occurs. That also is when the most weight transfer occurs. If we want the car to set in the middle of the turn, we might increase the right-side compression stiffness, reduce the left-side rebound stiffness—or, in some situations, both—to slow chassis roll. This delays the full onset of the weight transfer until the car is further into the turn.

Though changing shocks on the left or right sides of the car is most common, we also might change valving from front to rear to help corner entry or exit. If the car is pushing going in, we might encourage weight transfer to the front by reducing rebound (extension) stiffness at the rear shocks, or compression at the front. A car that is loose off of

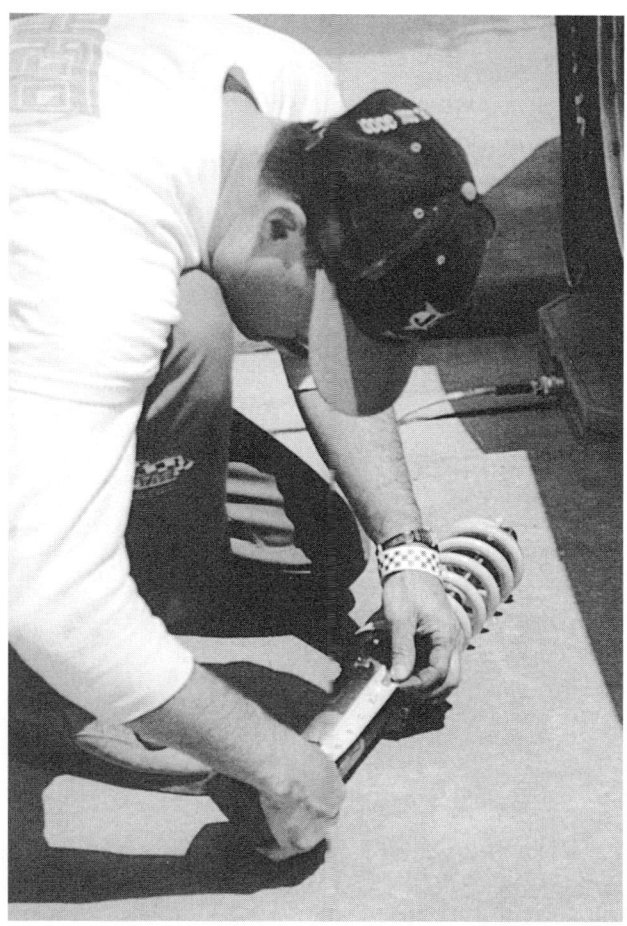

A crew member from Ray Cook's Hav-A-Tampa team makes careful shock changes while scaling the car. Because he is using the same spring, measuring the location of the adjusting nut helps get the setup very close before mounting the shock in the car.

the corner might need less compression stiffness at the rear shocks, or rebound stiffness at the front shocks to let the weight get back down on the rear of the chassis.

Keep in mind that there are a wide range of combinations that can be used. All four corners of the car must be looked at to see how one corner is affecting the others. This is a prime situation where good records can help keep you from getting lost. You have to know what you did and what the result was to stay on the right track. Otherwise you could soon be guessing and never get the car working or even back to where it was.

ROUGH TRACKS

When we get on a rough track, we may have to use stiffer springs to keep the car from bottoming out, but we may also use shocks with more rebound resistance to help prevent the car from jumping up and down so much as it passes through holes and

ruts. You can really glue the car down onto the track by increasing rebound valving. These valvings are more often used on the rear of the car, but some tracks and conditions may dictate increasing front rebound stiffness as well.

BANKED TRACKS

One of the more common changes we make on high-banked tracks is to increase the compression stiffness of the right-front shock and, occasionally, the left front as well. On very high banks, especially a very fast, high-banked track, we may have to increase the compression stiffness of all four shocks to keep the ride heights and suspension geometry from changing too suddenly when the car hits the banking. Using too much of the suspension travel too quickly, especially in the front, can allow geometry changes that normally help the car through the last part of the turn to occur before we want them. Slowing the speed at which the chassis sinks can help get the car setting at the correct point of the turn.

FLAT/SLICK TRACKS

On flat and/or slick tracks, we will usually soften up the compression and the rebound on the shocks considerably. In flat corners, the chassis does not sink as much, and on many tracks there may not be enough traction to cause the desired amount of body roll unless we encourage it with the shocks.

On some flat tracks we use 30/50 shocks (five compression and three rebound) on the front end to let the front of the car raise on acceleration. The result is more weight transfer to the rear tires sooner to help generate as much bite as possible to drive off the turn.

BRAKING FORCES

Braking forces acting on the rearend play a big role in how a car gets into the corner. In most cases, a single 90/10 shock is used to control the movement of the rearend as it responds to the driver getting off the gas or onto the brakes on corner entry. These shocks have a lot of resistance to extension, but very little on compression.

A 90/10 shock cushions the forward rotation of the rearend to prevent unloading the rear tires and loosening the car when the engine goes from driving to coasting, or decelerating on corner entry. On acceleration, the low-compression resistance lets the rearend respond to those forces normally.

This shock setup is particularly useful when the track starts drying out.

If the racing surface gets especially dry slick we may even use two 90/10 shocks. But when the track is very heavy we may not run any shocks at all on the rearend to prevent over-tightening the car on corner entry.

BASELINES

The easiest way to get a new car handling right is to listen to the chassis builder. They have a lot of experience with their cars and know what shocks will work in most situations. They might not be right-on-the-money for your driving style or track conditions, but they can almost always get you close. From there you can make small adjustments to the chassis and shocks to get the car more comfortable for the driver.

Shock manufacturers are also very capable of recommending a good shock setup for most chassis and track configurations. They have watched and talked with a huge number of racers, and know what is most likely to work in most situations. Here again, they might not be able to get the perfect setup under you, but they can get many teams closer and make the process of sorting the car out much easier and, ultimately, less expensive.

Getting close with shocks is pretty easy. Getting the car just right with shocks takes experimenting, common sense, and time. As with chassis adjustments, make small shock changes and keep track of what you did and what happened. As you get closer to the optimum setup, you may find other things on your car that need changing along with the shocks that will make you even faster. Handling is the offshoot of speed on dirt, and shocks are crucial to get any car working as best it can.

Notice the double-shear mounting at the top of the shock. This kind of mount and the brace running across the chassis between the tops of the shocks prevents the tubing from twisting and adding spring rate to that corner of the car. Double-shear mounting should be used whenever possible to resist breaking or twisting a single mounting tab out of the car. Tremendous forces are applied to the mounting, even on smooth tracks.

Chapter 9

SHOCK TUNING

Split Valves, Compression, Rebound—It Doesn't Mean Much If You Can't Get Power to the Ground

Text and Photography by Michael Thomas

In dirt Late Model racing the name of the game is all about planting the rear wheels to get forward traction.

Dirt racing is increasingly becoming a buy-it-and-bolt-it-on sport. One of the few ways to get an advantage over the next guy when it comes to the race car is good suspension tuning. To get the lowdown on getting a dirt Late Model hooked up with proper shock and spring selection, we went to Jeff Smith of J&J Racecars. In addition to being a darn good racer, Smith sells GRT and Warrior race cars from his shop in Gastonia, N.C., and spends a great deal of time helping racers find the perfect combination for their driving styles.

Circle Track: When tuning for a specific track, what areas of the car are the biggest factors?

Jeff Smith: It depends on how close we are on the setup when we get there. We generally have three areas we go to to tune. We go to the right-front spring, the Panhard bar or the top bar on the left rear (Smith runs a four-link rear suspension). We'll change the length of the bar or its mounting height to change the way the birdcage indexes. After that we'll get into shock valves. We'll run the same on the left front and the two rears. We'll run the same springs

there probably 90 percent of the time. What we change a lot is the right-front spring to affect the car getting in and off the corner. That's basically how we tune the GRT cars. I think most other cars are probably the same way.

CT: Like most dirt Late Model racers these days, you run a four-link rear suspension. Please explain the importance of running the shock behind the axle versus in front on the left rear.

Smith: Putting the shock behind the axle on the left rear gives the car more drive off the corner. It also makes it tighter. It does it because of the way the birdcage indexes, or moves, as the car rolls into the turn. As the car turns, it rolls over on its right side, picking up the left side. As the top bar on the left side moves down, it spins the birdcage around the axle. If the axle is mounted on the birdcage behind the axle it actually loads the spring, helping it push the left-side tire down to keep it in good contact with the track. The more pressure you put on that spring the harder it pushes the tire into the ground.

It's the opposite if you have the shock and spring in front

To get maximum traction on a four-link suspension, mounting the shock and spring on the birdcage behind the axle on the left side (top) and in front on the right (above) takes advantage of the bars to load the springs and press the wheels to the ground.

of the axle. As the left side rolls up and the top bar pulls the birdcage, it's actually unloading the spring. So if we are at a racetrack that's taking rubber and really getting hooked up, and you are able to be on the throttle all the time, the car will start rearing up with the shock behind and get too hard to drive. Then we'll go to the shock and spring in front of the rear end to calm the car back down.

The rule of thumb is you go in front of the axle tube on the left rear when the traction is there already. Also, putting the shock on the front of the axle will free the car up and keep it from rearing up so much. It's just reducing the traction available to the rear wheels.

CT: *Do you switch types of shocks depending on track conditions?*
Smith: I run both monotube and twin-tube shocks. For consistency, I stick with just one brand of shocks and springs, Afco in my case.

CT: *What factors determine what type of shock you will run and where?*
Smith: If I'm going to a racetrack that I've been to several times and I know what I need setup-wise, I'll put my regular shocks on, which are Afco twin-tube units. I like the standard shock the best because I feel they react a little quicker than a gas-charged shock (monotube). But that's just my seat-of-the-pants feel.

If I'm going to a racetrack I'm not familiar with and I'm not really sure what I'm going to need, I'll put on a set of double-adjustable shocks. That way I can adjust both compression and rebound individually and not have to be slinging a bunch of shocks and springs around. It's a big advantage to be able to tune your shocks without having to chuck a bunch of units out of the rack. These shocks are also twin-tube, and they are neat because they allow you to adjust in any increment you want, not just whole numbers.

Monotube shocks are more consistent on rough tracks, so I'll switch to the gas-charged shocks if I think the track is going to get rough. With the constant up and down of the wheel, you don't run the risk of foaming the oil with a gas-charged shock.

CT: *As a driver, can you tell the difference between a standard twin-tube shock and one that offers adjustability?*
Smith: Not really. Lots of times on my double adjustables, I'll set the compression and rebound a half-pound softer than I will run my regular shocks. If I normally go on 75s and 94s, on my double adjustables I'll put the settings at 74.5 and 93.5. I'm

not saying that's the right thing for everybody, but it seems to work for me in terms of getting the same feel from the driver's seat.

CT: Ever mix shock types?

Smith: I've played with some monotube stuff on the right rear, then loaded up on the gas pressure just to try to get the car back on the left rear quicker. I've mostly just experimented with things like that. It's got some potential in some situations.

CT: Some chassis builders now allow you to change rear linkages quickly. Does switching from, say, a four-link to a Z-link require a change in valving?

Smith: Usually, no. Switching the linkage configuration is mainly for controlling rear steer, especially when going from a four-link to the Z-link. You are just flipping the top bar from the front to the back. Now when the car rolls over it's keeping the rear end in the same place up and down. The only reason you would change your shock valving in this situation is driver preference.

CT: Give us a guideline of some situations where shock valving is a good tuning option.

Smith: Every situation is different, so it's difficult to give hard and fast rules. But a good example is say your car is a little loose getting in and across the middle of the corner. You think it's because the car isn't planting its right side hard enough. What you can do is go to your left-side shocks and soften up the rebound a little so that side can rise up faster. Then soften up the compression on the right so it will drop faster. You can go the opposite direction if the car is too tight—stiffen the compression on the right and the rebound on the left. You may not have to adjust both sides. Sometimes you can go to just one corner of the car. For example, the left rear has a much greater influence on the car coming off the corner than it does coming in.

CT: What is the one piece of advice you most often find yourself giving to inexperienced drivers and chassis tuners?

Smith: I guess the thing I get most often with people who are still learning is they see what somebody else is running and they hear about things like split valves and soft compressions, and automatically they think they've got to have the same stuff. They may be way off somewhere else on the race car, but suddenly it's the shock that's supposed to fix all the problems. It's tempting to think you can fix about any handling problem with shocks and springs, but you are only covering up a

The shock mounts to the birdcage on the left side behind the axle. The birdcage allows the axle tube to remain stationary (controlled by the pullbar) while still indexing the shock and spring.

problem, not fixing it. If somebody tells me they need a certain shock combo, my first question is always, "Why do you think you need it?" The right answer is never because that's what somebody else is running. You've got to be looking to solve the problems on your car, not somebody else's.

CT: Now for the fun one. What's the next evolutionary step in dirt Late Model suspension tuning?

Smith: Some people have begun playing with double shocks on the right rear, one in front of the axle and one behind. The idea is to have zero compression with rebound valving on one and zero rebound with compression valving on the other. Generally, with the double springs you will run half on each spring to total what you would with a single shock-and-spring setup. But then, some people do it a little differently, say if you normally run a 200-pound spring there—instead of going to a pair of 100s, some people will use something like a 110 and a 125. So it isn't a hard rule that you put even weight springs on this configuration.

We haven't done enough yet with the double shock/spring setup to know exactly what we need with it. I've tried it a little at some test sessions and found things that I thought were really good. But then when I got to a race and tried to apply what I had learned, the car didn't respond the way I had expected, and it wasn't what I needed.

The double-spring setup seems to be a good idea for people running with a spec tire rule. But then when they've gone to an open-tire race, they've always struggled. I've even struggled trying to run the thing in an open-tire race, because when you can use any tire compound you can get the car so hooked up you are going to run tight, and the double-spring setup makes the car tighter. On a spec tire rule you are not as tight and the setup will help you out. But again, as far as exactly what spring rates and valving work best, we haven't quite figured that out yet. It will just take some time to figure it out unless something else comes down the pipe before then.

Chapter 10

DIRT TRACK CARBURETION

Five Carb Adjustments for More Horsepower

by Michael Thomas
Photography courtesy Demon Carburetion

Once you've done your homework and spent your hard-earned money on a carburetor and fuel system for your dirt track car, it's time to tune the system for optimum power. Of course, everything changes with weather, altitude, and track conditions, so your carburetor settings that worked real well last week may not be what you need this time around. Getting optimum performance out of your carburetor and fuel system, no matter what the conditions, will make dirt track racing a lot more fun, and hopefully, profitable too.

1. CARBURETOR SELECTION

Just like a good set of cylinder heads and manifold, a well-prepared carburetor with improved airflow and corrected fuel metering throughout the rev range will do wonders for an engine's performance. There is, though, a definite science to carburetor preparation, and the old adage, "The bigger the better," is not always true.

Small carburetors tend to provide better throttle response at lower engine speeds, while bigger carburetors are inclined to make more power in the mid- and upper-rev ranges. When running on a dry, slick racetrack, a large carburetor that produces less power at lower rpm may be the ideal solution for keeping the car hooked up when exiting the corners. Conversely, on a tacky racetrack a smaller

carburetor with lots of low-end power will provide strong acceleration off the corners.

When the track is sticky, some racers will qualify with a smaller carburetor and switch to a larger model if the track becomes slick. If you are restricted to only one carburetor, it's advisable to have it blueprinted and flowed to obtain maximum performance.

2. IDLE ADJUSTMENTS

Instant part-throttle response is mandatory for a successful dirt track car, and idle-mixture adjustment plays an important role in achieving it. Many racers seem to feel that if they take their foot off the accelerator pedal and the engine doesn't quit, the car is idling fine. The truth is that initial acceleration, throttle response, and smooth deceleration are all related to a properly balanced idle system.

On most carburetors, the fuel controlled by the idle-mixture screws is discharged from one or more holes in the baseplate below the butterflies. This provides sufficient fuel to run the engine when the butterflies are virtually closed. As the throttle opens, the butterflies uncover the transfer slots, which provide fuel for the initial surge of air entering the manifold. Carburetors not responding to adjustments to the idle-mixture screws usually have too much of the

Adjust the idle-mixture screws so that the engine idles properly with the butterflies closed.

One of the more innovative carburetors on the market is the Demon, which permits interchange of booster and venturi components that make adjusting to track conditions quick and easy.

transfer slots exposed at idle.

If this situation exists, the transfer slots are probably already open, and no fuel is available to mix with the incoming air until the discharge from the accelerator pump arrives. This condition is chiefly responsible for off-idle stumbles and hesitation. When a carburetor is correctly calibrated for idle, the engine will run with the transfer slots closed, and a smooth transition from closed to fully open throttle is ensured.

Correct idle calibration will also help steady engine performance during deceleration. The carburetor is exposed to extremely high manifold vacuum during deceleration, and if the butterflies cannot fully close, excessive fuel can be drawn from the transfer slots. This causes flooding and backfiring. Flames from the exhaust during deceleration are a good indicator that the idle circuit is not properly calibrated, and the butterflies are not restricting fuel flow from the transfer slots.

3. FLOAT LEVEL

Correct float-level adjustment helps guarantee that your carburetor will neither flood nor run out of fuel. It's also difficult to obtain a good idle setting if the float levels are incorrect. One of the easiest ways to check float level is to remove the float bowl from the carburetor, turn it upside down, and measure the gap between the float and the bowl. The gap should measure 4/10 inch.

To check float levels on carburetors without sight glasses, remove the float-level sight plug while the engine is idling. Take care that fuel does not spill on the hot engine and create a fire hazard. Fuel should barely trickle from the primary or front end of the carburetor and should be slightly higher at the rear. The hexagonal nut on the needle and seat will adjust the float. To adjust the needle and seat, loosen the screw in the middle of the assembly slightly, and retighten when adjustments have been completed.

4. POWER VALVE ADJUSTMENT

The power valve provides a way of leaning the fuel mixture to the engine under low- and no-load conditions. Power valves are rated in inches of vacuum and are numbered accordingly. When the engine manifold vacuum drops beneath the number stated, the power valve opens and enriches the main circuit. To keep the idle clean and sharp, the power valve should remain closed during idling. To check them, use a vacuum gauge and read the engine's vacuum (at idle).

The power-valve number should read 1.5 to two inches of vacuum under the engine's manifold vacuum. For example, if an engine idles at eight inches of manifold vacuum, a 6.5 power valve would be appropriate. The power valve will not open and enrich the circuit until the engine vacuum drops to 6.5—as it decreases toward zero vacuum when the throttle is opened to or near full throttle. This ensures the power valve will remain closed at idle, keeping the engine clean and the spark plugs from fouling.

5. GETTING THE RIGHT MIXTURE

Maximum horsepower is essential for winning races, and the correct fuel mixture is vital for making maximum horsepower. Proper jetting is one of the major tools used to accomplish this. Fuel, oxygen, and a heat source are the three components necessary for combustion. The heat source (the ignition system) ignites the fuel and oxygen mixture to create energy. The amount of energy that can be produced is based on the amount of fuel that is burned. Of course, the amount of fuel that can be burned is dependent upon how much oxygen is available.

Unfortunately, the chain of variables doesn't end there. The amount of oxygen available is determined by the density, or weight, of the air. The three factors that influence air density are temperature, barometric pressure, and humidity. Lower temperature, higher barometric pressure, and lower humidity all increase the density of air. As air density increases, higher oxygen levels allow more fuel to be burned, and power output increases. As air density decreases, so does engine power. This is why lap times often show substantial improvements after the sun sets and the coolness of the evening sets in.

Racers and engine tuners can monitor these changes in air density and use the information to calculate fuel-mixture adjustments. They should also be aware of chassis or driveline changes that must be made to compensate for changing power levels. To predict the effects changing weather has on your engine, use quality instruments to measure temperature, barometric pressure, and humidity. Those readings can then be fed into one of the many devices on the market that will evaluate your inputs and return optimum fuel-mixture settings. Taking full advantage of all the information available to you provides a competitive edge and greatly increases the odds of visiting the winner's circle. To be successful, a dirt track carburetor must be tuned for the conditions at hand. Here's hoping these tuning tips enhance your chances of winning.

Chapter 11
REAR STEER DESIGN

How to Make Rear Steer Work For You

Text and Illustrations by Bob Bolles

A 4-bar dirt Late Model rear suspension is designed to have a large range of rear steer. The adjustability allows the racer the opportunity to make adjustments for changing conditions that occur on dirt surfaces. The attitude of the car on dry, slick tracks can be quite radical.

Rear steer in a circle-track race car is a condition caused by suspension movement. Under the right conditions, rear steer can be beneficial and enhance performance. Under the wrong conditions, it can ruin your handling.

We do not necessarily need to know the exact amounts of rear steer our cars are subjected to, but we do need to have a solid understanding of what produces rear steer and what effect rear steer has on the handling in our cars.

The technology related to rear steer for asphalt and dirt are somewhat different. There are a few similarities, but many differences in how we evaluate and use rear steer for each group. Since we cannot lump them together, we will analyze them separately.

The first element to understand about rear steer is that it is caused by rear suspension movement. As the rear corners of the car move, along with the controlling arms that locate the rear-end fore and aft, each side can move the wheel on that side forward or to the rear. Obviously, if both of the wheels did not move, or moved in the same direction by the same amount, we would have zero rear steer. When one wheel moves more than the other, we have a certain amount of rear steer.

Rear steer can either tighten a car or make it very loose. Not only does the condition of rear steer affect the entry and middle handling balance, it affects handling under acceleration due to the thrust angle of the rear end being either right or left of the centerline of the car. In dirt racing,

it may be advantageous to incorporate a large amount of rear steer under certain conditions.

TUNING WITH REAR STEER

To see if rear steer is an important tuning tool, we turned to some experts. Joe Garrison of GRT Race Cars and Mark Richards of Rocket Chassis both said having rear steer capability in a dirt car was critical. "Rear steer helps the driver get the rear end around on dry, slick tracks without having to break the rear tires loose," said Garrison. Richards added that "if the driver has to countersteer the car a lot, you need more rear steer."

Keith Masters of Masterbilt Race Car Chassis said, "The reason chassis builders build 4-bar cars is because of rear steer. The entry is more important to design for than the exit as far as rear steer is concerned." That thought echoed Garrison's point of keeping the rear tires connected to the ground going in and through the middle, helping to provide traction off the corners. A smooth entry provides a faster exit.

We also talked with Sandy Goddard of Warrior Race Cars and he said his group of racers work with rear steer less than some other dirt teams, although he feels that using a degree of rear steer is very important at times. "Rear steer definitely helps get into the corners on dry, slick tracks," Goddard said, "but you can put too much rear steer into the rear suspension. A lot of guys take that to extremes."

The front mounting block on the 3-link suspension system is slotted vertically so you can adjust the angle of the trailing arm to fine-tune the amount of rear steer. The car is very sensitive to changes in the trailing arm angle and the height of the front mount should be moved in small increments.

C. J. Rayburn of Rayburn Race Cars, a driver as well as a car builder, echoed much of what the other builders had to say. "We have always wanted rear steer," he said. "Rear steer is important in any design of dirt car." Rayburn builds the swing-arm type of rear suspension. Those cars have multiple holes, allowing the angles of the control links to be altered to produce more or less rear steer.

A rear end steered to the left of centerline will cause the thrust angle to be left of centerline and make the car tighter on entry and tighter on exit under acceleration. A rear end that is steered to the right of centerline causes the thrust angle to be pointed to the right of centerline and makes the car loose on entry and loose on exit under acceleration. Knowing these basics, we need to look at each type of racing to see how rear steer affects each type of car and how we might improve our performance using rear steer.

ASPHALT REAR STEER

The asphalt racing surface provides a lot of traction, even on those flat "slick" tracks. Because there is very little slip of the tires on asphalt, the range of useable rear steer is very small. We never need our suspension to steer the rear end to the right of centerline on asphalt. It has been a practice for teams to align the rear end—and/or have it steer slightly to the right—to fix a tight mid-corner

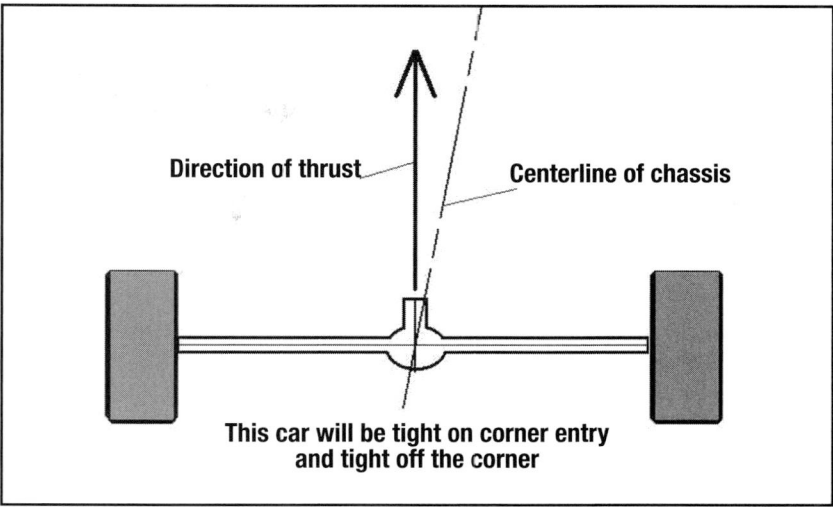

This car will be tight on corner entry and tight off the corner

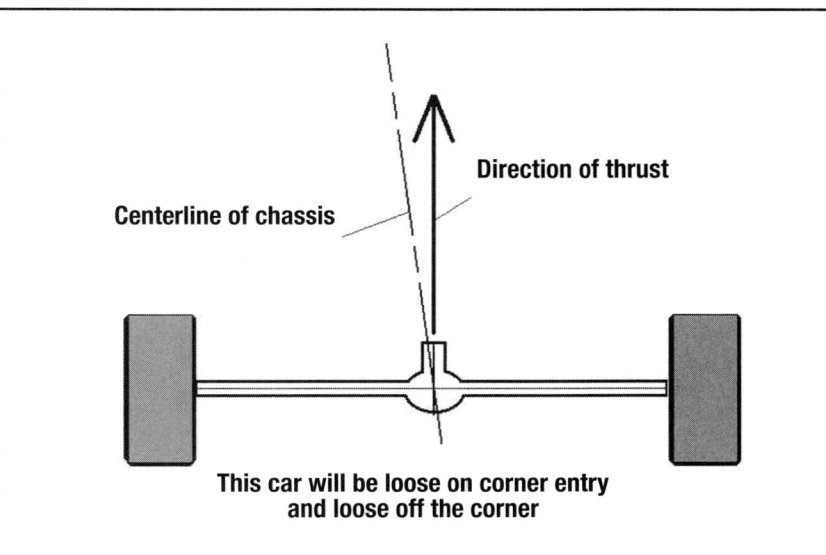

This car will be loose on corner entry and loose off the corner

With the rear end pointed to the left of centerline, the steering will cause the rear of the car to want to run under the front end, causing a very tight condition, especially under acceleration. With the rear end pointed to the right of centerline, the car will be freed up going in, through the middle, and possibly loose off the corners, with the rear end wanting to run around to the right of the front.

condition, but this goes into the category of crutches, and should not be necessary if the car is set up properly otherwise.

Asphalt stock cars have four predominant rear suspension systems and all of them produce some amount of rear steer. They are:

1. The Asphalt 3-Link

The 3-link rear suspension system has two trailing arms mounted near the rear tires, and one third link mounted atop the rear differential that controls rear end wrap-up. The trailing arms can be mounted parallel to the centerline of the car, or angled with the front mounts closer to centerline.

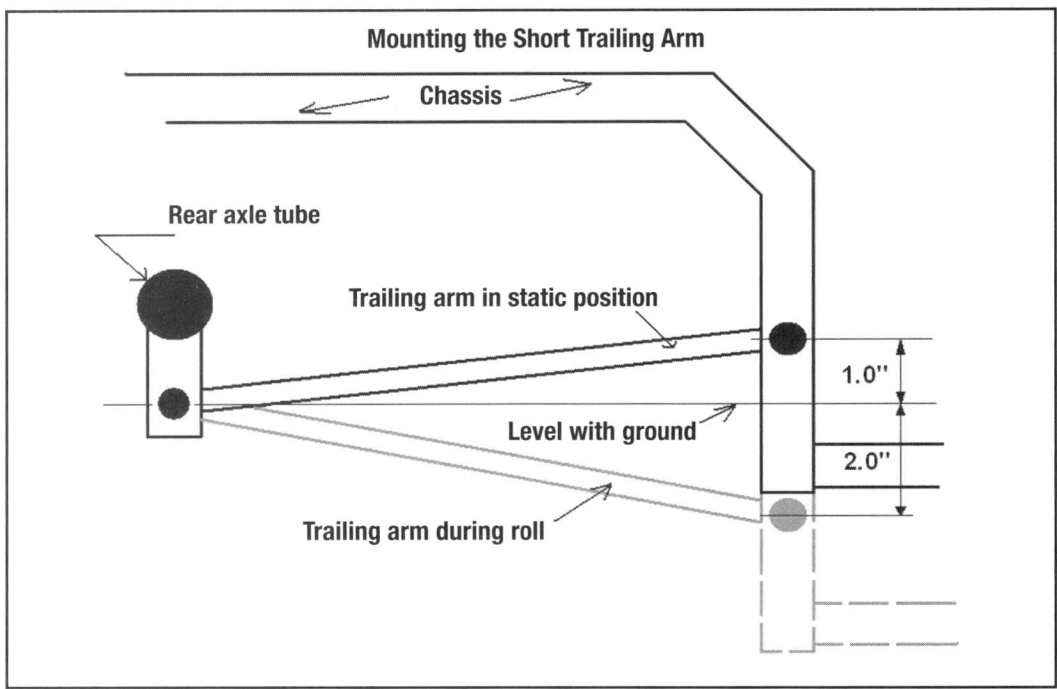

Mounting the Short Trailing Arm

Chassis

Rear axle tube

Trailing arm in static position

1.0"

Level with ground

2.0"

Trailing arm during roll

The 3-link rear suspension system can produce rear steer in both directions. As the chassis moves down on the right side, the right-rear wheel will be moved back as the front mount approaches the height of the rear mounting point. As the front mounting point continues to move down, the right-rear wheel will be pulled forward. Normally, to produce a small amount of rear steer to the left on asphalt, we mount the front pivot point 1/3 of the total travel distance higher than the rear pivot.

The pull bar, upper-third link allows the rear end to rotate under acceleration. The movement of the rear end rearward can be utilized to produce rear steer to the left only while under acceleration. That way, the car is correctly aligned at the rear during entry and through the middle of the turns. Rear steer to the left helps promote bite on flat and slick asphalt tracks.

Rear steer in this system is caused by chassis movement, which can produce several secondary effects. Usually, the right-rear corner of the chassis moves more than the left-rear. On most flat- to medium-banked tracks, the left-rear moves very little. This has been confirmed by studying data from onboard computer systems during testing that show shock travel amounts in the turns. The left-rear shock mostly seems to move between 1/2-inch

in rebound and up to 1/2-inch in compression during the entire lap. The right-rear shock shows from 3 to 4 inches or more of travel, depending on the spring rates used.

On most asphalt 3-link cars, the right-rear trailing arm mostly controls rear steer due to body roll. We usually need to position the angle of the trailing arm so that the front mount is higher than the rear mount by roughly one third of the distance that the front mount will move down during cornering. The variation of height for the right-rear trailing arm is very small. Changes in the height of the front of the trailing arm as small as a 1/4-inch can be felt by the driver.

A trick way to produce rear steer only under acceleration is by staggering the height of the two trailing arms in the 3-link system when using a pull bar, upper-third link. If we mount the left-side trailing arm lower than the right-side trailing arm, then as the rear end rotates under acceleration due to the pull bar extending, the left-rear wheel will move rearward more so than the right-rear wheel, causing rear steer to the left to a small degree. This promotes forward bite without causing the car to be tight on entry or in the middle of the turns.

Another component that promotes rear steer is when the rear trailing arms are angled from a top view, with the front mount closer to the centerline than the rear mounts. With this design, lateral movement of the rear end causes rear steer. If the Panhard bar is mounted on the right side of the chassis and level to the ground, the rear end will be pulled to the right, and will steer to the left when the chassis moves during cornering. This is caused by the rear end swinging around the instant center, created from projecting lines through the arms to the front until the lines meet.

2. The Truck-Arm System

The truck-arm system has been adapted from the design for a 1964 Chevy truck, and is used on many Late Model Stock cars, as well as the three premier divisions of NASCAR (Craftsman Trucks, Busch, and Nextel Cup). These systems only steer to the left and have a limited amount of steer. The

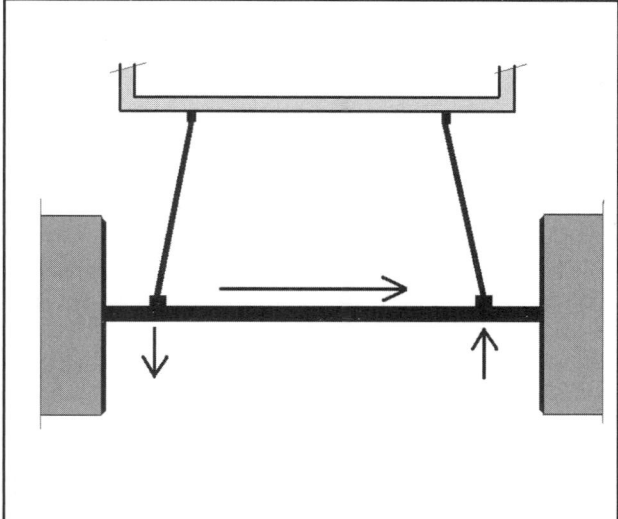

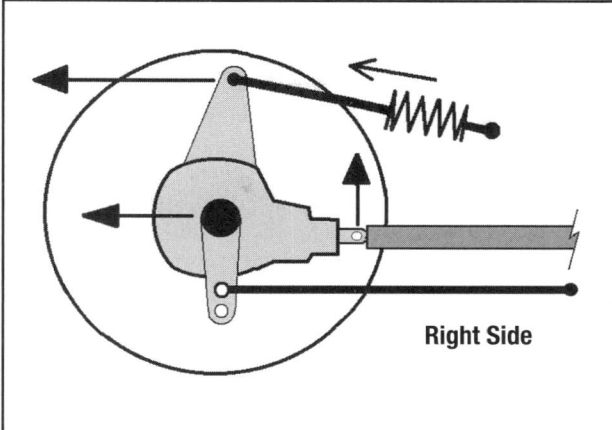

Right Side

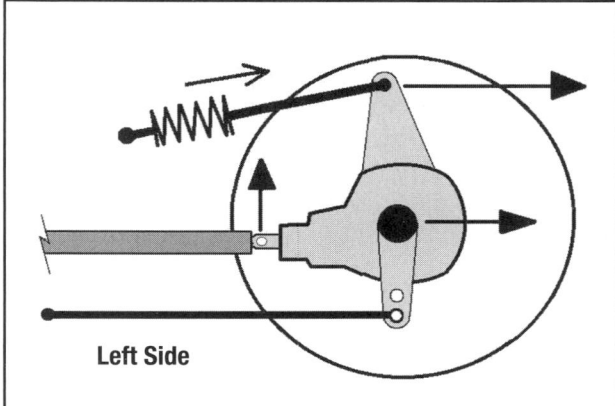

Left Side

As the pull bar extends under acceleration, the rear end rotates back, causing the rear wheels to move to the rear. The bottom pivot point is the back end of the trailing arms. If we mount the right-side arm higher than the left-side arm, the radius on the left side will be longer from the lower mount to the axle, causing the left-rear wheel to move farther rearward than the right-rear wheel. This causes a slight amount of rear steer to the left, the exact amount depending on the difference in height of the mounting points and the amount of third-link travel.

Some 3-link rear suspensions are built with the trailing arms angled from a top view with the front mounts located closer to the car's centerline than the rear mounts. This creates some amount of rear steer if the rear end moves laterally as the chassis rolls. This type of rear steer will tighten the car if the rear end moves to the right as the car rolls. The downside to this rear steer is that it will tighten the car in the middle of the turns, as well as off the corner.

roll of the chassis and the movement of the Panhard bar are the two components that influence the amount of steer in these systems.

As far as the geometry related to rear steer is concerned, this is an ideal system. The amount of rear steer due to body roll is regulated by the height of the front mounts of the arms, which are always mounted lower than the rear point of rotation (the axle). Rear steer amounts—due to the Panhard bar angle—are regulated by the angle. A downside to using the truck arms is not related to steering characteristics, but due to a narrow spring base when the springs are mounted directly on top of the truck arms, creating a narrow spring base in the rear.

3. The Metric 4-Link System

The metric 4-link is a widely used system that comes with some models of stock automobiles. It uses four links, as the name implies, that are not parallel to the centerline of the car. The top links are angled from a top view with the front pivots wider than the rear pivots. The lower links are angled from a top view with the front pivots narrower than the rear pivots.

With this system, the rear end stays located by virtue of the opposing angles of the upper and lower links. There is also steer to the left using this system, and because of the width of the front mounts of the lower-controlling links, rear steer can

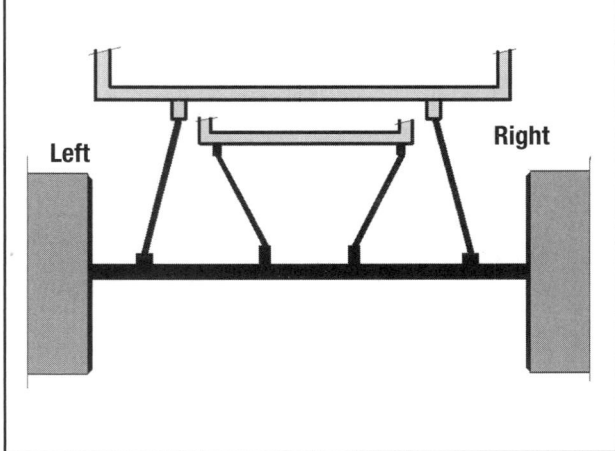

Left   Right

The metric 4-link suspension has two links above the rear end and two links below the rear end. They are angled from a top view to prevent the rear end from moving side to side as the chassis rolls.

be considerable. Under most current rules, there is no adjustment for amounts of rear steer with these systems.

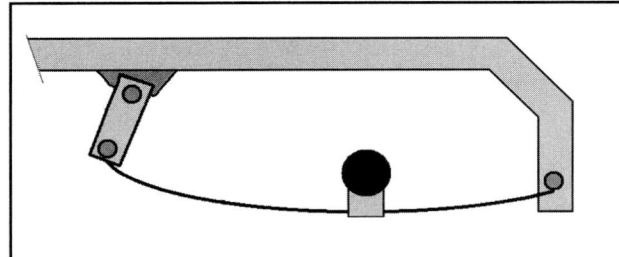

The leaf-spring suspension system usually has fixed mounts in the front and therefore no adjustment for amount of rear steer. The rear end stays mostly square in the car throughout the lap. The feedback we hear from the racers is that the leaf-spring rear suspension is good on tacky or wet tracks.

4. Leaf-Spring Systems

The leaf-spring rear suspension system locates the rear-end fore and aft, as well as laterally using the leafs. There can be a small amount of rear steer as the chassis rolls and squats, but it is both minimal and mostly fixed as far as adjustability. The advantage of this system is that it keeps the rear end squared up and the thrust under acceleration straight ahead, if that is what is needed.

DIRT REAR STEER

There are four types of rear suspensions used in most dirt cars that are significant to study regarding rear steer. Characteristics of the metric 4-link, one

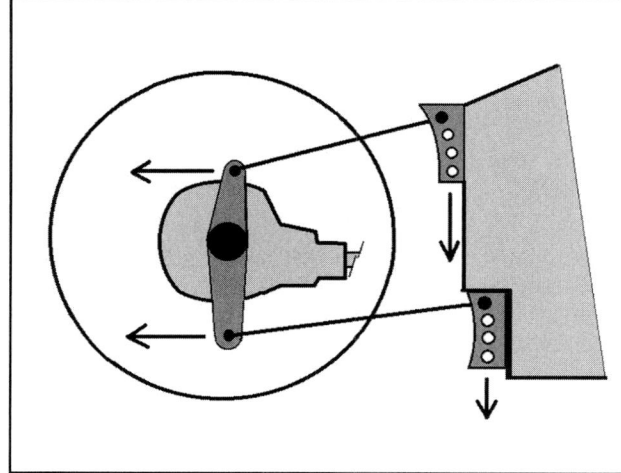

If the bars on a 4-bar car are set all the way to the top of the mounts, the rear end will steer to the right quite a bit as the car rolls up on the left side and down on the right side. This helps the driver get the car turned to prepare to exit the corner, and keeps the rear tires in contact with the track surface under extremely dry, slick conditions.

of those four, are the same as discussed under the heading related to asphalt cars. Let's expand on the other three systems.

Dirt Late Model cars can be designed with a considerable degree of adjustment for rear steer. Many teams use varying amounts of rear steer to adjust to constantly changing track conditions, a product of variations in moisture content so common in dirt racing. Other teams may just stick with a fixed location for the mounts in the rear end and adjust handling with other means.

One of the reasons a car will test fast in practice, qualifying, and maybe the heat races, but then be out to lunch when the track changes come feature time, is a result of improper rear steer. Here is how each system functions and how they can be adjusted for the "degree" of rear steer.

The Standard 4-Bar System

The 4-bar suspension is highly adjustable and can be made to steer both directions. The rule about never steering the rear end to the right on an asphalt car does not apply on a dirt car. There are times when we definitely want the rear to steer to the right.

Depending on the angles of the trailing arms or bars, each rear wheel can be made to move to the front or rear. The roll angles and vertical movement on a dirt car can be very pronounced. With so much movement, we can plan out our rear steer just about any way we need it.

The bars can be mounted on one side of the car so that only that wheel moves to create the rear steer. If both sides are configured to move in opposite directions, then rear steer can be extreme.

On a tacky track, the team would do well to limit rear steer on both sides of the car. These conditions call for a driving line that is more straight ahead. When the track goes slick, especially dry slick, rear steer is needed. In the past, drivers would set up the car for exit off the corners, throwing the car sideways by breaking the rear tires loose. In more recent years, teams have been setting up the car so that the left side raises up quite a bit.

The left-rear suspension is designed so that when that corner raises up, the arms are angled, pulling the left-rear wheel forward toward the driver. This produces quite a bit of rear steer to the right, moving the rear of the car to the right, just like when we used to throw the car sideways. The difference is that now we can maintain rear traction—having never broken loose—and the car is angled somewhat sideways, and pointed in the right direction to get off the corner.

How Much is Too Much?

There are limits to how far we go in steering the car this way. One disadvantage is pointed out by Masters. "High left-rear loading does not increase traction," he said.

As the left side of the car travels up, the front of both of the trailing arms are angled upward so that the left-rear tire tries to drive up under the chassis, loading the left-rear tire considerably. We can have too much weight end up on the left-rear tire and lose traction and/or cause the car to push off the corners because all of the forward thrust is concentrated in the left-rear tire. In racing, we have the maximum amount of traction from a pair of equally loaded tires on the same axle. Excess loading of either the left- or right-rear tires decreases traction in most cases.

DIRT TRACK AERODYNAMICS

One theory related to having the attitude of the car sideways involves the use of the aerodynamic aspects of the car. If you look at a modern dirt Late Model, the sides are made up of big, flat panels similar to the sides of a Sprint Car wing. We can see the effect of the Sprint Car wing in the turns—as the cars actually roll left due to the pressure differential developed on the flat wing sides—as the cars go sideways at a high speed. On a Late Model, this air pressure difference may help keep the car on the track by virtue of having the car go through the air at an angle, causing both a high pressure on the leading (right) side and a lower pressure on the trailing (left) side. Masters added, "Any time we can press air against the sides of the car, we can help the car to turn." His company has experimented with dirt track aero in the past.

The reason rear steer aero might be important comes on dry, slick surfaces, where the tires do not grip well. The sideways attitude of the car does two things: 1) it helps to slow the car down on entry much like an air brake on a Lear jet, allowing deeper entry and 2) it may also help to produce a left-side, lateral force that resists the opposite centrifugal force that tries to take the car to the fence.

THE Z-LINK SYSTEM

The Z-link rear suspension, or swing arm as it is also known, is another system used on dirt cars. Compared to the 4-bar cars, it has more limited adjustment for rear steer and historically has worked well on the tighter and more highly banked race tracks because the rear end is pointed more straight ahead. Some manufacturers have added

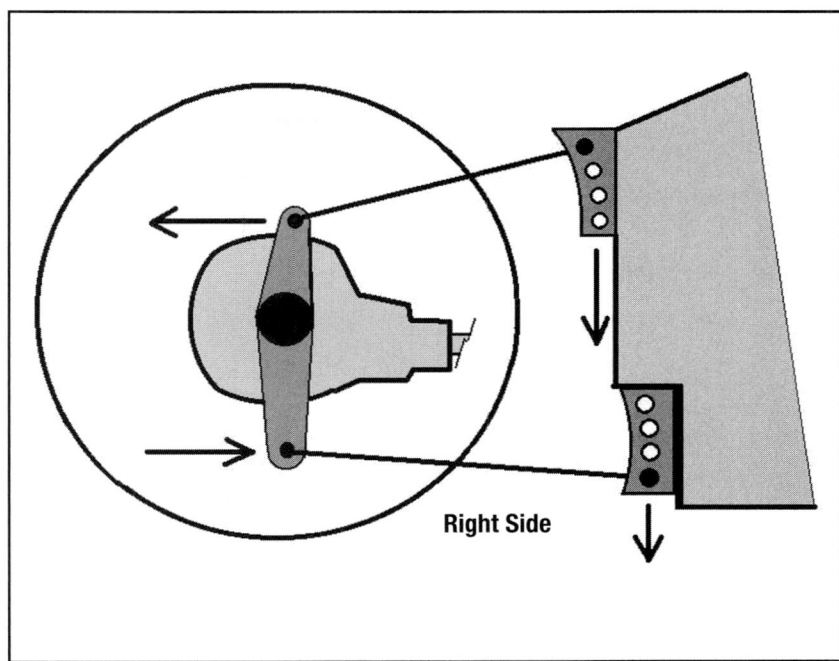

Right Side

If the bars on a 4-bar car are set in the correct holes, the movement of the top and bottom mounts at the rear end will compensate fore and aft, resulting in zero rear steer. This is best for tight, wet tracks where we cannot use any rear steer and we need for the thrust to be pointed straight ahead.

multiple mounting points on the front and rear chassis mounts. This helps make the rear steer characteristics more adjustable for the changing conditions. Richards added that today the Z-link or swing-arm suspension can have nearly as much rear steer as the 4-link, without the excessive loading of the left-rear tire.

Spring Motion Ratio

Most of the Z-link systems utilize a spring mounting system that attaches the coilover spring to the front link. This produces a motion ratio that causes the spring to move less than the chassis per degree of roll and/or inch of squat. Therefore, the rate the car feels is much less, usually around 50 percent, than the actual installed spring rate. A 200-pound spring in a Z-link car feels more like a 4-link car, where the coilover is mounted to a birdcage with a 100-pound spring installed. The significance of this, for the purpose of this article, is that the chassis travel in a Z-link is enhanced compared to the 4-link suspension when using the same installed spring rate, and this causes quite a bit of chassis travel and related rear steer. So, teams need to take this rate difference into account.

There are reports from the past of a team winning a race using four 400-pound springs on a Z-link type of car. The front of the car felt the actual 400-pound rate while the rear "felt" like 200 pound

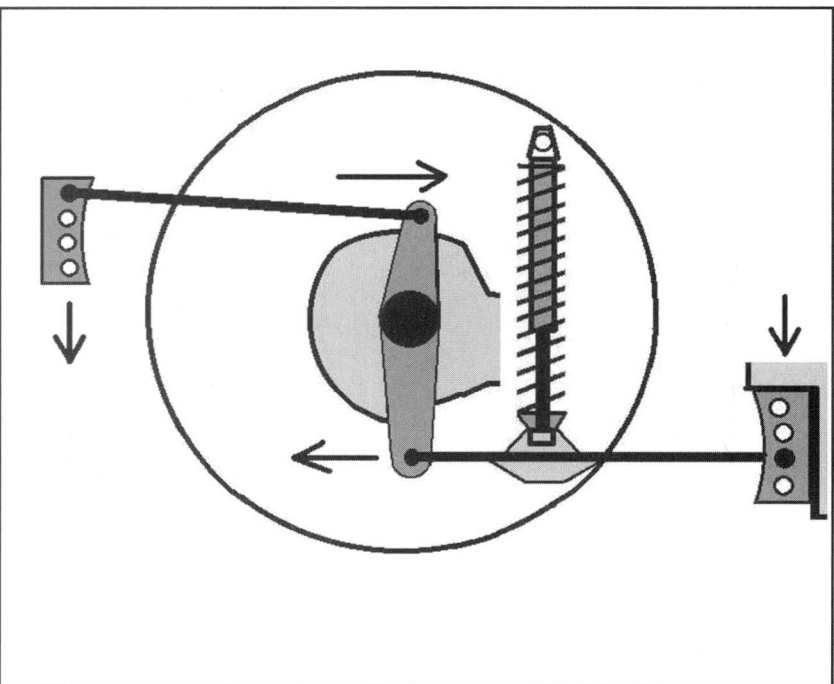

A Z-link suspension system uses a link extending from the rear end forward to the chassis, and one from the rear end rearward to a mount on the chassis. Most designs use very few mounting holes that would enable the team to adjust for the amount of rear steer in those cars. Note the spring is mounted directly on the front link.

springs. The track in this case was banked and had a lot of grip. The car was set up and driven more like an asphalt car and it was fast. Those conditions rarely exist on dirt.

Tuning With Rear Steer

We should learn to read the conditions of the dirt tracks and tune the amount of rear steer—less for tacky and wet conditions, with more rear steer as the track gets slicker. On extremely dry, slick conditions, use lots of rear steer to the right. This is accomplished by causing the right-rear wheel to move back and the left-rear wheel to move forward as the car rolls. Soft springs, a left chassis-mounted track bar, and easy-up shocks on the left side all promote the body roll that produces rear steer to the right.

On asphalt, do not make large changes to components that influence rear steer. Make small adjustments if you feel a need. When you find the correct amount of rear steer, stay there and tune the handling with the other components. When racing on dirt, watch the conditions and be prepared to make changes accordingly, not just to the setup, but also related to rear steer. That way, the car will stay as fast and balanced as it can be throughout all of the changing conditions.

DIRT CHASSIS
SETUP TIPS

Chapter 12
IMCA PRE-RACE PREP

Text and Photography by Bob Ryder

The proper pre-race preparation of a chassis setup at the shop will help put you ahead of the competition on race day.

In Chapter 1, Jim Doffing, owner of Flexi-Flyer, gave us an introduction to what makes an IMCA-type Dirt Modified work. In this issue, we will cover the pre-race preparation and set up of an IMCA Dirt Modified at the shop.

CHASSIS SET-UP AT THE SHOP

The main goal for setting up the chassis at the shop is to have the car ready to race competitively as soon as you roll off the trailer. If the car is set up at the shop correctly, you should have to make only a few minor adjustments at the track.

Find a flat, level surface in your shop to use as your setup area. Always use the same place. Mark on the floor where the tire patches are located, so the car can be relocated every time for setup.

You should do your chassis setup in a specific order every time. The car should be completely race-ready with correct springs and all of the fluids full. The wheels, tires (including tire stagger), and air pressure should be the same as if you are at the track ready to take the green flag. Don't forget to add the correct amount of ballast in the driver's seat to simulate the driver's weight.

The following chassis-adjustment procedures should be in this order:
1. Set ride-height at each corner.
2. Set the desired weight distribution.
3. Set crossweight.
4. Set caster.
5. Set camber.
6. Set toe-out.
7. Front roll centers.
8. Rear roll centers.
9. Spring rates.
10. Recommended basic starting baseline specs.

We have already talked about setting ride-height. The next step is to get the car on the scales. Make sure the scales are calibrated and zeroed-out to ensure their accuracy. After you get the car up on the scales, disconnect the shocks and antiroll bar (if you are using one) to make sure that the car's true weight is not held up off of the springs.

Adjusting the weight distribution of the car is accomplished by adjusting the corner heights of the car. A small amount of adjustment should be necessary if the weight masses have been properly positioned. If more than one or two turns on the weight-jacking screw is required at a corner, work on two jack screws at a time. For example, if more rear weight is desired, lower the left-front and right-front jack screws one turn each, then raise the left-rear and right-rear screws one turn each. If more crossweight is required, lower the left-front and right-rear screws one turn each, and raise the right-front and left-rear jack screws one turn each.

Example:

Your car has a finished race-ready weight of 2400 pounds, including the driver and 20 gallons of fuel. Your final desired

This mock-up chassis sits without a full body for easy inspection of suspension components. A powdercoated chassis sure does make for an easy cleanup.

weight is 52.5-percent left weight, 53.5-percent rear weight, and 49.5-percent crossweight.

Now we add up the numbers and find we have achieved 52.5-percent left weight and 53.5-percent rear weight, but the crossweight (right-front and left-rear added together and divided by the total weight) equals 50.2 percent, which is more than desired.

The crossweight has to be set by screwing down on the right-front and left-rear weight jackers. You will find that no matter what crossweight you use, the left-side weights will always total the same, and the rear weights will always total the same. The total weight mass of the car sets the left-to-right and front-to-rear weights. The weight jackers affect the diagonal weights.

To figure the desired corner weights that will yield the target crossweight perform the following:

1. Multiply the total weight of the car (2400 pounds) by the desired crossweight percentage (49.5 percent), which equals 1188 pounds.

2400 lbs. x 49.5% = 1188 lbs.

2. Subtract the target crossweight, or 1188 pounds, from 1204 pounds (total of the crossweight), which equals 16 pounds.

1204 lbs. - 1188 lbs. = 16 lbs.

3. Divide 16 pounds by 2, which equals 8 pounds.

16 lbs. ÷ 2 = 8 lbs.

4. Subtract 8 pounds from the right-front and left-rear corners by screwing up equally on the weight jackers at those corners. If you have to change the corner height at the left rear and right front by more than 1/2 inch, screw down equally on the left-front and right-rear weight jackers until the proper crossweight is achieved. After we have done this, we found that the subtracted weight at the right front had to be added to the left-front weight, and the subtracted weight at the left rear was added to the right-rear weight.

SUGGESTED WEIGHT Distribution

The following is the recommended starting baseline weight distribution for dirt track cars on flat- to semi-banked tracks.

Left side = 52 to 53 percent
Rear = 53 to 54 percent
Crossweight = 49 to 53 percent

These percentages are assuming a normal track width of 64 inches in the front and 60 inches in the rear.

Running on an average dirt surface (that's firm

With body panels removed, rear suspension components, driveline, brake lines, fuel lines, filter, and other parts become very accessible.

The front suspension consists of '70 Chevelle lower control arms, Afco upper A-arms, Afco springs, Afco screw jacks, Afco shocks, a GM '72 station wagon brake caliper with Wilwood pads, '72 station wagon heavy-duty hubs and rotors.

and has good moisture), the crossweight will fall between 49 and 50 percent and is usually 49.7 percent. But since each jockey rides a horse differently, we must consider the driver, his feel for the car, and how dry the track surface is. A heavy, wet track will require less crossweight to minimize bite and push; on the other hand a dry, slick track will require more crossweight to get more bite.

If you find yourself with more than 53 percent of left-side weight on a flat- to semi-banked dirt track surface, you are going to have trouble getting side bite at the right rear, and the car will have a tendency to be loose going into the corner.

These Modifieds on dirt don't like a lot of left-side weight. Also the cars are not effective with a lot of rear weight. More than 53-percent rear weight will have a tendency to cause a push going into the corner and coming out. But on a "stop and go"–type track, which consists of two dragstrips connected with two tight turns at each end, the car should be 30- to 40-pounds left-rear heavy because there is so much instantaneous weight transfer on corner entry. Running on a fairly round track, where the car keeps up its momentum in the corners, or on a higher-banked track (more than 12-degrees banking), the car should be 30- to 40-pounds right-rear heavy to prevent push. On a very dry, slick track, add 40 pounds of ballast to the very rear of the car to enable more rear bite.

Crossweight

Crossweight is also interpreted as diagonal weight. It is the total of the right-front and left-rear corner weights divided by the car's total weight.

More crossweight adds more bite or understeer into the chassis. More crossweight is used mostly on paved or dry, slick tracks to keep the rearend of the car tight on corner entry, and to improve bite off the corners. It favors the left-rear tire contact patch by heavily loading it.

To compute the crossweight, add together the single-wheel weights of the left-rear and right-front corners. Then divide this number by the total vehicle weight. Your answer will be a percentage, which is the crossweight percentage.

Car Total Weight = 2400 pounds
Corner Car Weights:
   Right-front = 530 pounds
   Left-rear = 647 pounds

647 pounds left-rear weight + 530 pounds right-front weight = 1177
1177 -:- 2400 = 49.5%

Whether your car weighs 2400 pounds or 3200 pounds, the crossweight relates to the same amount of percentage weight set diagonally in the car. Find the right diagonal-weight percentage that works best for you, such as 51 percent or 55 percent, for example. If you do change the total weight of your car, you can come back to your baseline setup.

Crossweight and tire stagger work hand-in-hand with each other. Crossweight puts bite in the car for better entry and corner exit but it can occasionally cause the car to push or understeer. Now adding more tire stagger will help balance the chassis. A common rule is that the more crossweight a car has, the more stagger it needs.

FRONT-END ALIGNMENT
Caster

The caster should be done before the camber because camber is affected by caster. Caster will provide directional steering stability. This is created by a line that is projected from the steering pivot axis down to the ground. This line contacts the ground out in front of the tire contact patch when the caster is set in a positive position. The torque arm is between the projected steering axis pivot line and the center of the tire contact patch. The torque arm serves to force the wheel in a straight-ahead direction. The longer the length of the torque (caused by greater amounts of positive caster), the greater the steering effort required to turn the wheels away from this straight-ahead direction.

A difference in caster setting between the left-front tire and right-front tire is called "caster stagger." Just a slight amount of caster stagger helps the car change from its straight-ahead path more easily to ease into a left turn—which is what oval racin' is all about (when the caster at the right-front is greater than the caster at the left-front).

Caster stagger on a dirt track car generally is not very large because the same factor that aids in turning the car to the left increases steering efforts when turning to the right to countersteer. Caster and caster stagger used on a race car is also influenced by the use of power steering. A car with power steering can use more positive caster and more caster stagger. We found out that the Chevelle chassis with the angled swing axis of the A-arms will develop a little more than 0.5 degrees of caster gain per inch of travel.

Jim Doffing was kind enough to inform us that most IMCA Modifieds use the following guidelines for the initial caster settings

Dirt track car (manual steering)
Left-front: + 1.5 degrees
Right-front: + 3 degrees

These turntables from Reb-Co make it easier to check caster. They can be locked to 0 degrees and used in conjunction with scales, which makes scaling and front-end adjustments quick and easy.

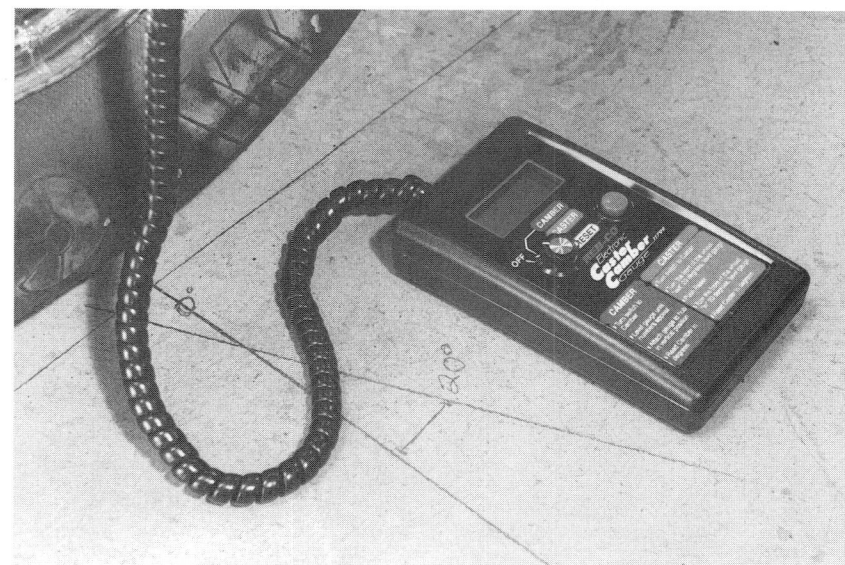

Dirt track car (power steering)
Left-front: + 2.5 degrees
Right-front: + 5 degrees

Camber

Camber adjustment is to keep the tire contact patch flat on the track surface at the maximum point of cornering. Camber is the biggest contributor to the vehicle's cornering ability. With IMCA Modified cars, the static camber setting at the right-front tire is between 2 and 3 degrees negative, depending on the camber change curve of the suspension, the type of track and banking, and

The above is in contrast to the old school of checking caster and camber on the floor with a straightedge and protractor. Reb-Co also has a caster/camber electronic digital gauge that takes the guesswork out of dialing-in the front end or rearend.

the tire construction.

The following guidelines are for the initial camber settings:

Dirt track car
Left-front: + 1 degree
Right-front: – 2.5 to 3 degrees

The amount of body roll and the front roll-center height all dictate the amount of static negative camber at the right-front. This car has a front roll-center height of 3 inches, it will not need as much right-front static negative camber as a car with a lower roll center of, say, 2 1/2-inches, because the higher front roll center gives the car more negative camber gain. When there is less initial negative camber, it helps a car in transition into a corner because the right-front will have more tire footprint on the track at this critical section of the corner.

Toe-Out

The amount of total toe-out that is correct for your car depends on the car's front-track width, amount of Ackerman steer, and the turn radius of the racetrack. The basic guideline for toe-out is 1/8-inch to 3/16-inch out for both dirt and asphalt. The tighter the turn radius, the more toe-out is required. Tire temperatures taken after some practice laps will help you dial-in the exact toe-out required. For an inexperienced driver, using a 1/4-inch toe-out might be advisable. This will put a slight push in the front end to tighten up the car. This is a more stable feeling for the inexperienced driver.

Reb-Co also has made it a lot easier to check bump steer because the unit's frame is leveled on its three screw pads. A hub plate is mounted and held by three lug nuts. Zero-out the feeler gauge on the hub plate. Make sure the feeler gauge and pointer on the other frame arm are level and perpendicular, find the zero-point of suspension travel, and begin plotting bump steer.

Chapter 13

IMCA TRACK TUNING

**Text by Bob Ryder and Jim Doffing of
Flexi-Flyer Industries
Photography by Bob Ryder**

When the chassis is tuned and dialed-in, the driver feels that he or she becomes part of the car. The driver is able to set the chassis entering the corner, continue through the apex, and carry the maximum power and speed coming off the corner and down the straight.

In the last chapter, Jim Doffing, owner of Flexi-Flyer and an IMCA-Modified race car manufacturer, gave us this advice: "Use the same shop floor position every time, and mark the four tire patches so setup is repeated the same every time." Also, the preparation procedure is conducted in the same sequence every time.

This chapter will help you to understand the major principles of IMCA-type chassis tuning, including recommended baseline specs chassis adjustments at the track, corrections at the track, fine-tuning with tire stagger, and wheel offsets.

ROLL CENTERS

There are two roll centers that work in conjunction with the front and rear suspensions. The line connecting these two points is called the roll axis. This axis is the location where the body rolls (leans) when the race car goes through a corner, the body leans in the opposite direction that the car is turning. This transfers the balanced weight to the right side of the chassis, which creates torque to the right side. The torque can be resisted or controlled by springs and sway bars.

Front Roll Centers

The front suspension and steering play major roles in the car's handling characteristics. The two most critical design considerations in the front end of the car are the roll-center placement and camber-change curve. The roll-center placement includes both vertical and horizontal locations.

The true roll center of the front independent suspension is that point where the instant center swing arms of the left and right intersect each other.

To get the whole picture, we must consider the left-front and right-front roll centers together with what is happening with the front end. In an oval track car you are only turning left. The left-front corner is moving as well as the right-front during body roll and cornering. Doffing tells us it is best to keep the roll center as close to the centerline of the car as possible, moving the roll center away from the centerline will cause a leverage effect on the chassis; that is undesirable.

On an IMCA dirt Modified, the correct design for the front roll center is 2.75 inches and 3 inches above the ground. This measurement would be acceptable for a car running on flat to moderately banked tracks. If you are racing on high-banked tracks the car will require a lower roll center, making the camber change curve less.

These dirt Modifieds don't respond well to high front roll centers. The main problem is the tires will not generate enough lateral acceleration to get the tires to generate side-bite, which gets you off the corners quicker. So it must be acquired by using body roll. To accomplish this, a lower

BASIC IMCA SPECS

Track type: Flat- to medium-banked 3⁄8-mile
Car weight: 2400 lbs
Weight distribution: 52% left, 53% rear, 49.5% crossweight

Monoleaf/coil rear suspension spring rates:
Left-Front: 650 lbs   Right-Front: 750 lbs
Left-Rear: 200 lbs   Right-Rear: 200 lbs

Shocks:
Left-Front: 75   Right-Front: 76
Left-Rear: 94   Left-Rear: 94

Front-end alignment:
Caster (manual steering):
Left-Front: +1.5°   Right-Front: +3°

Caster (power steering):
Left-Front: +2.5°   Right-Front: +5°

Camber:
Left-Front: +1°
Right-Front: -2.5 to -3°

Toe
Toe-out: 1⁄8"

Ride-heights:
Left-Front: 4.5"   Right-Front: 5"
Left-Rear: 5.25"   Right-Rear: 5.5"

Track adjustments:

If you have done your proper preparation at the shop, you should only have to make minor adjustments at the track.

Both front and rear spring rates are what makes the car hook up. Getting the correct spring rate, along with the right roll-center heights will create more body roll, which creates more side-bite getting the car to hook up and get off the corner.

front roll center and softer spring rates should be used.

Jim says, the camber-curve change of a front suspension is directly tied to the roll-center height. The "lower" the front roll center, the "less" camber change there is per inch of wheel travel. The "higher" the front roll center, the "more" camber change there is per inch of wheel travel.

Also, mass or weight is the major factor of front roll center. Less front mass requires a lower front roll center. A larger and higher front mass requires a higher front roll center. For example, if we are comparing a Late Model Sportsman car to a lighter weight chassis like an IMCA Modified, the Late Model Sportsman needs a higher front roll center. This is because the Late Model has a heavier mass up front over the front wheels; a Modified has a greater engine setback.

Another big influence on body roll center is the type of tire being used. The IMCA Modifieds use a very hard tire. The harder the compound the lower the roll center needs to be to create more downforce and bite on the tire. A lower roll center creates more body roll, and combined with lighter spring rates, creates the side-bite that these narrow, hard tires require.

Every driver drives his or her car differently, some are erratic, and some are smooth as silk. A car with a higher roll center is a more forgiving car. A car with a lower roll center will be less forgiving because the erratic loadings and unloadings of the chassis will be amplified.

Rear Roll Centers

Adjusting rear roll centers will affect the weight transfer from the left to the right. If the car uses a Panhard bar or leaf springs (monoleaf or multileaf) in the rear, it is very easy to move the rear roll center up or down.

Dirt Track	
(Leaf-Spring Rear Suspension)	
750 lbs left-front	900 lbs right-front
95 lbs left-rear 9	5 lbs right-rear

Dirt Track	
(Monoleaf/Coil Rear Suspension)	
650 lbs left-front	750 lbs right-front
150 lbs + 50 lbs left-rear	150 lbs + 50 lbs right-rear
(Monoleaf)	(Monoleaf)

Dirt Track	
(Stock Chevelle Four-Link Rear Suspension)	
850 lbs left-front	900 lbs right-front
175 lbs left-rear	150 lbs right-rear

The torque-rod (arrow) baseline should be at 20 degrees; increasing the angle of the torque bar will hook up the rearend harder. If you start setting it at 25 to 30 degrees, wheel hop can occur and can unhook the car down the straightaway. The pinion angle should be 7 to 9 degrees.

Lowering the rear roll center increases side-bite at the right rear (on dirt). The rear roll center can be used to adjust the amount of side-bite. If the rear roll center is too high, and the front springs are too stiff, the car will go into a four-wheel drift. The same happens if the roll centers are high at both the front and rear. If we lower the rear roll center, we will achieve more side-bite, and probably cause a push. Now we can go to a softer spring rate in the right front to balance the car. The overall effect: a little more body roll can create side-bite.

Rear roll center can also be used to adjust a push out of the car. To do this we move the Panhard bar up, which raises the roll center and takes away bite at the right-rear, thereby eliminating the push. Also, if the right-rear spring rate is too stiff (causing a loose condition), and we don't have the proper spring rate available, we can lower the rear roll center a little to improve the handling.

The same technique can be used to adjust a car that races on a dirt track that starts out real wet and ends up real dry. For a wet track condition, start with the Panhard bar higher, then move it down as the track dries out.

SPRING RATES

The secret to getting the IMCA Modified tire to hook up is a relatively soft spring rate. These softer spring rates, along with the appropriate roll-center heights, create more body roll and that creates more side-bite.

The front spring rates used depend on the lower A-arm being used. The '78 to '86 GM mid-size car's lower A-arm has a spring pocket that is more inset toward the center of the car; this exerts more leverage on the spring, and a stiffer spring has to be used than would be used with the '73 to '77

The rear roll center is controlled by the Panhard bar ("J" bar)—lowering the bar will increase the side-bite at the right rear (on dirt). Raising the Panhard bar will raise the roll center, taking away the right-rear bite (causing a looser condition).

Chevelle lower A-arm.

The rear spring rates depend on the type of rear suspension system being used, and the rear roll-center height. IMCA Modifieds use those darn small, hard tires and to get them to work and get some side-bite, the car has to have a lot of body roll. Spring rates and roll-center height go hand-in-hand with getting the correct amount of body roll. A suspension system that has a high roll center, such as a stock Chevelle four-link system, requires very soft spring rates in order to generate enough body roll. Lower roll centers, like having a 2-inch lowering block on a monoleaf, which gives you about 9 inches, or roll center, requires stiffer springs to control the roll.

IMCA Modifieds respond well to a positive spring split in the front, the right-front spring rate

A chassis with push (understeer) means you are turning the front wheels left and the car still goes straight. The rear roll center is too low; by raising the Panhard bar ("J" bar) you are raising the roll center. This takes the bite out of the right rear and eliminates the push.

The car may go into a four-wheel drift at the apex or middle of a corner, if the rear roll center is too high, and the front springs are too stiff. Or also, if the roll centers are too high in both the front and rear.

is stiffer than the left-front. In the rear, even spring rates, or slightly stiffer left-rear spring rates work well. Remember, cars that run real heavy left-rear weights will usually use "straight-up" front springs—the same spring rate at the left-front and right-front. They will usually use a stiffer right-rear than left-rear spring.

Doffing uses the following guidelines for basic "ballpark" starting spring setup for an average dirt track that is a firm-packed, 3/8-mile, 0- to 10-degrees banking and has a good amount of moisture. This is with our 3-inch front roll center. The car also has a monoleaf/coil spring rear suspension that has a 9-inch rear roll center. If you are running a three-link coil spring, the rear

suspension will have about a 12-inch rear roll-center height.

Springs are like people; they change with age and use. When you first install new springs, use a good spring-rate checker to double check the true rating, and also measure the free height of the springs. Enter these figures in your chassis log book, so you will have a baseline to compare against in the future. Occasionally, take the springs out, then rate and measure them to see if there have been any changes. Jim Doffing says they have rated the springs on many race cars, and have found that the actual spring rate is as much as 25-percent softer than the racer had thought.

TORQUE BISCUIT ROD

Remember to set the rubber biscuit preload after the chassis heights and weights are set. The starting angle of the biscuit rod is 20 degrees. More angle, like 25 to 30 degrees, will hook up rear tires harder, but wheel hop can occur and can also unhook the tires down the straightaway. This preload sets pinion angle; make sure you check it. Recommended pinion angle is 7 to 9 degrees. Jim has found that heavier springs might be needed in the rear because more downforce is created.

When sorting out chassis problems you must do it in an orderly manner. Let's say you are having trouble with turn-in entry. This can affect the car's handling all the way through the turn, and getting off it. When figuring this problem always start at the braking point into the corner, then follow up with analyzing turn-in entry, corner apex, and corner exit. The car should be reviewed at each of these phases, and in that order.

Condition: Push on Entry
1. Too much front brake bias.
2. Too much front spring.
3. Too much front shock.
4. Rear springs too soft.
5. Rearend too far to the left (change wheel offset).
6. Gear too high.
7. Right-front spring too soft, and it bottoms out.

Condition: Push at Apex
1. Rear roll center too low—raise.
2. Acceleration push at rear: the rear steer is wrong; shorten the right side.
3. Right-side shocks too stiff.
4. Sway bar too heavy or preload too heavy.
5. Tire stagger not enough, add more rear stagger.

Condition: Push at Exit
1. Too much crossweight.

Changing the wheel offsets will change the handling characteristics of the car. For different track conditions, change offsets. (Bob Ryder Photo)

2. Rear steer too long on right side.
3. Increase tire stagger.
4. Left-rear spring too stiff.
5. Gear too tall.
6. Possible offset wheel add 1/2-inch wheel spacer to right-rear and left-front.

Condition: Wheel Drift (Usually occurs at apex or middle of corner)

1. Right-side air pressure too high.
2. Right-side springs too stiff.
3. Left-side weight too high.
4. Too low rear roll center.
5. Too low rear center of gravity.
6. Front and rear track wrong. Move rearend to the right using offset wheels or move front end to the left using offset wheels.
7. If it occurs under direct acceleration, trail is too long or stagger is wrong.

Condition: Wheel Spin (Weight transfer is not occurring)

1. Not enough rear weight.
2. Too much rear gear.
3. Torque biscuit rod needs more angle (anti-squat).
4. Torque biscuit rod rubbers too tight or too stiff.
5. Rear springs too stiff.

Example/Review

After hot-lapping, you have found the car has developed a push going into the corners. What should you do to correct it, and in what order should you correct it? Try using these adjustments in order; remember to just make one change in the chassis at a time!

1. Use more rear stagger.

2. Take preload out of the front antiroll bar.

3. Take some crossweight out of the chassis. Remember to adjust each corner of the car to keep the ride-heights balanced. Screw down equally on the left-front and right-rear, and screw up equally on the right-front and left-rear. Never make more than a full-turn adjustment in either direction on either front weight jack bolt. More than a turn at a time will dramatically affect the front suspension geometry.

4. If the problem is more severe, and the previous adjustments don't cure the problem, then a spring-rate change should be made. To do this, decrease the right-front spring rate to the next lowest spring rate (work in 50-pound increments at the front corners and 25-pound increments at the rear corners). Fire the car up and test these changes.

5. If you find the problem still exists, there are other areas to evaluate. First, check the tire temperatures, and see if there is an indication of a front-end alignment problem—possibly not enough negative camber on the right-front. Also, check the average temperature for the left-front tire. If it is too cool, move some ballast up the left-side framerail to put more physical mass at the left-front corner. But before you move the ballast, be sure to measure the corner heights of the chassis so the same balance can be restored after the ballast is moved.

6. Another area to look at is rear-roll understeer. This is caused by the rear-suspension linkages moving the right-rear tire ahead and the left-rear back during body roll. A good indication of this to the driver is that the understeer increases as the body roll increases. The understeer will be proportional to the amount of body roll because it is the body roll that is causing it. To cure this problem, reset the linkage lengths or front attachment heights so that the rearend is not steered by the linkage movement.

7. Does the understeer occur under braking during corner entry? Then the problem could be too much

FLEXI-FLYER TRACK TEST

These are the Flexi-Flyer baseline conditions:

Left-Front
Wheel offset: 4 inches
Tire circ.: 78 3/4 inches
Tire pressure: 10 lbs

Right-Front
Wheel offset: 2 inches
Tire circ.: 79 1/2 inches
Tire pressure: 10 lbs

Left-Rear
Wheel offset: 2 inches
Tire circ.: 79 1/8 inches
Tire pressure: 8 lbs

Right-Rear
Wheel offset: 4 inches
Tire circ.: 80 3/4 inches
Tire pressure: 9 lbs

With the car set up at this baseline we found the car to be to tight after the first session; it needed to be loosened up, so we changed only the stagger.

Left-Front
Wheel offset: 4 inches
Tire circ.: 78 3/4 inches
(changed)
Tire pressure: 10 lbs

Right-Front
Wheel offset: 4 inches
Tire circ.: 79 7/8 inches
(changed)
Tire pressure: 10 lbs

Left-Rear
Wheel offset: 2 inch
Tire circ.: 79 1/4 inch
Tire pressure: 8 lbs

Right-Rear
Wheel offset: 4 inch
Tire circ.: 81 1/4 inch
Tire pressure: 12 lbs

BASIC WHAT-TO-DO CHART

Car Loose Into Turn:

Item	Adjustment
Stagger:	Decrease right rear
Toe-out:	Not enough
Front springs:	Too soft
Rear springs:	Too stiff
Antiroll bar:	Too soft
Brake bias:	Too much rear
Suspension bind:	Exists in rear suspension
Crossweight:	Increase

Car Pushes Off Corner:

Item	Adjustment
Stagger:	Increase right rear
Toe-out:	Decrease
Front springs:	Too stiff
Rear springs:	Too soft
Antiroll bar:	Too stiff
Crossweight:	Decrease

This is a very basic quick reference guide to common handling problems. These references are to give you an idea of where to start when you experience a problem.

front brake bias. Adjust the brake proportioning bar to move more braking bias to the rear.

If your car is loose instead of pushing, use the same sorting procedure as we just talked about, but just make the adjustments opposite of what was suggested to cure the push condition.

For example, first try decreasing rear stagger, then add some preload to the front antiroll bar, followed by adding crossweight to the chassis. If you find you need a spring-rate change, your first change should be to increase the right-front rate.

CHASSIS-TUNE WITH WHEEL OFFSETS

To find the basic setup for an average dirt track (firmly packed dirt with some moisture) use a 3-inch wheel offset at three corners, and a 2-inch offset on the right-front.

Now, if you find a very wet track condition, you need to move the right-rear out to decrease bite. Here you would use a 2-inch offset at the right-rear.

If the car still pushes, then move the rear in by using a 4-inch offset at the left-rear. For a dry, slick track, more bite is needed at the left-rear and right-rear. To achieve this, pull the right-rear in and kick out the left-rear. Use a 4-inch offset at the right-rear and a 2-inch at the left-rear. You can use a 1/2-inch wheel spacer at the left-rear to move it out even farther if needed.

If the car is still loose after making these changes, add 40 pounds of ballast to the very rear of the car to get more rear bite.

After the changes were made only to tire stagger, the results were better, but still a little tight. The car seemed to be fine in turns one and two. But in turns three and four the car just didn't want to loosen up in the middle of the corner; it still had a little push. Sometimes you can't please the car's handling from one corner to the other, and a trade-off or compromise must be made.

Chapter 14
IN SEARCH OF THE ULTIMATE SETUP

How the Pros Track Test Their Suspension Setup

by Bob Bolles and Will Handzel
Photography by Will Handzel

Editor's Note: Although this test took place nearly a decade ago, we're presenting it here because the basic test methods, and indeed a lot of the technology, are just as pertinent today. The test occurred at Eldora Speedway, and included top drivers and teams, product manufacturers and engineers at the time.

For the drivers and cars, it was decided only the best would do. Between them, Billy Moyer, backed by his GVS Racing crew, and Kevin Weaver, with the Eaglin Motorsports Racing team, have more than 700 career wins.

So, if these guys aren't at the top, we don't know who is. Both teams brought two cars to test multiple setups and brought their complete rigs with plenty of spare parts. From the moment the cars were unloaded to late into the night, the wrenches and parts were flying.

To complete the test team, there were company representatives on hand from Bilstein Shocks (DeWayne Ragland), Hyperco springs (Kelly Falls), Pi Research (Mike O'Gara, Mark Chambers, and Dave McIntosh), and Chassis R&D (Bob Bolles). The shock and spring guys were there for specific product testing, and the Pi Data boys showed up to instrument the cars so that we could gather a bunch of on-track data. Bob Bolles from Chassis R&D did a lot of consultation with the race teams before and during the test to get the cars consistently fast. As a note, Bolles and Ragland were the driving force behind getting this test to happen.

Besides the generic optimization of the suspension setups, it was decided before heading to the track that we would test a new pull bar being developed by Bert and Hyperco (the pull bar mounts on the rear-end housing to control

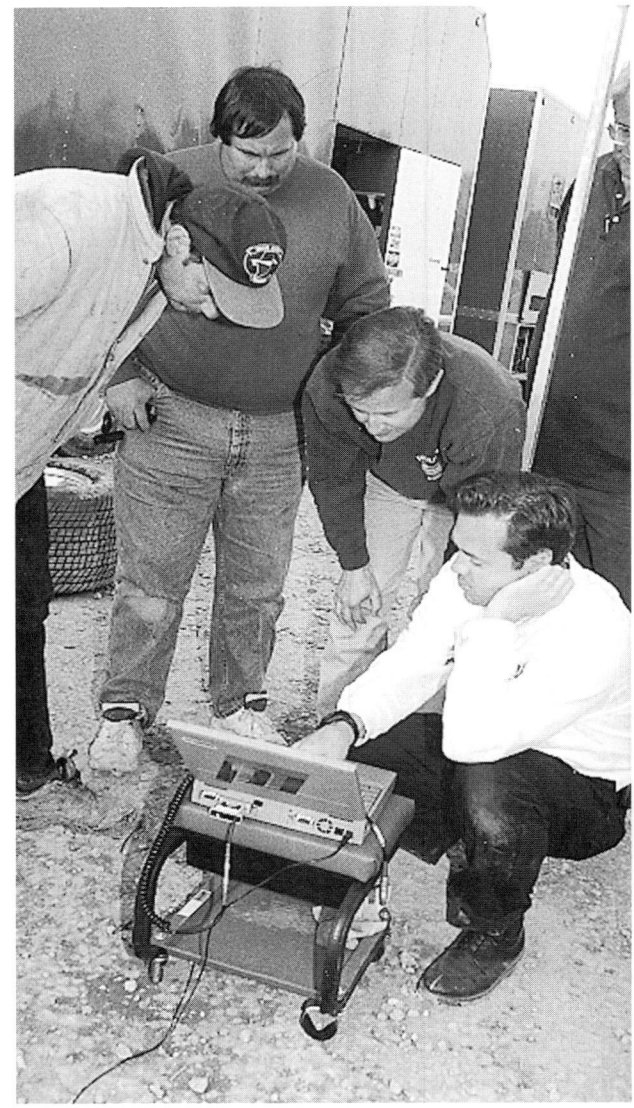

axle wrap under acceleration or deceleration) and determine if different-length springs make a difference in the handling of a car for Hyperco. The Pi Research system would be used to record lap times, segment times, throttle position, engine rpm, vehicle speed, suspension displacement at all four corners, rear-end wrapup, and more if possible.

The general suspension setup goals included improving how the car turned in the middle of the corner, improving the forward bite of the rear tires when exiting the corner, and improving the consistency of the laps times and general performance.

Determining the optimal suspension setup on dirt is difficult at best. Actually, doing any scientific testing at a dirt track has more impossible in it than difficulty. This is because the track surface can go from muck to tacky flat to dust in under an hour of high-noon Midwest sun, and the driving style on these surfaces can change so dramatically that the suspension-setup data gathered means nothing. The people involved in this test thought long and hard about this and decided to leave the track surface the way it was after the previous race; a race was held at Eldora only a few days prior. The track was relatively smooth, had some moisture in it (but not much), and was starting to take some rubber in the groove. While this wasn't necessarily a fast track, it was a consistent surface, even if

While most track testing is done using the stopwatch as the only indicator of an improvement or loss, for this test it was felt more data needed to be recorded, especially on the ever-changing surface of a dirt track. A PI Research System 3 was used (see installation in photo A). With this system, we recorded the throttle position (photo B), wheel speed (right-front versus right-rear), g-force, brake application, shock travel and velocity (photo C), and rear-end wrapup (photo D). When laid out in graph form, in relation to a fully segmented track-timing layout, we could easily see where gains or losses were happening. The PI systems range in price from $1,000 on up, but the information generated by them is truly valuable in determining what the car is really doing.

it was a few tenths off the pace. The day of the test was slightly overcast, which worked in our favor—keeping the track predictable all day. This allowed us to learn some important lessons.

The first part of the testing was in getting the cars to turn well. While it might look cool, dirt racers lose a lot of time sliding the rear of the car around on the entry to a corner to get the car through the turns. Selecting the right suspension components (springs, shocks, and so on) and setting them up so that the car can get through the corner as quickly as possible is the goal.

Once some baseline laps were in the books, the racers started to discuss possible changes in setup with the specialists on hand and referred to the Chassis R&D software recommendations before changing springs, shocks, and other components. Only one change was made at a time, then the driver was sent back out onto the track to gather data. As a side note, everybody at the track can vouch that Moyer and Weaver were hanging it out

on every lap, brushing the wall in a few instances, but delivered repeatable lap times in the process. So, the data gathered was a true ten-tenths view of the race car's performance potential.

In an effort to get suspension geometry optimized, the racers consulted with Chassis R&D's Bolles before the testing. The program is very effective in getting the front and rear roll center set so that they allow the car to work freely through the corners.

The first changes made to the cars were to soften up the front springs. Everyone agreed that this should help the cars in the entry and middle of the corner. This was done because a common problem among dirt cars is that they run too stiff of a spring in the right-front. The result is that the race car lifts the left-front tire off the ground as the car travels through the corner, effectively wasting the traction potential of that tire. If the tire isn't completely in the air, but is being unloaded, this problem can be detected in tire temperatures, or in the case with the

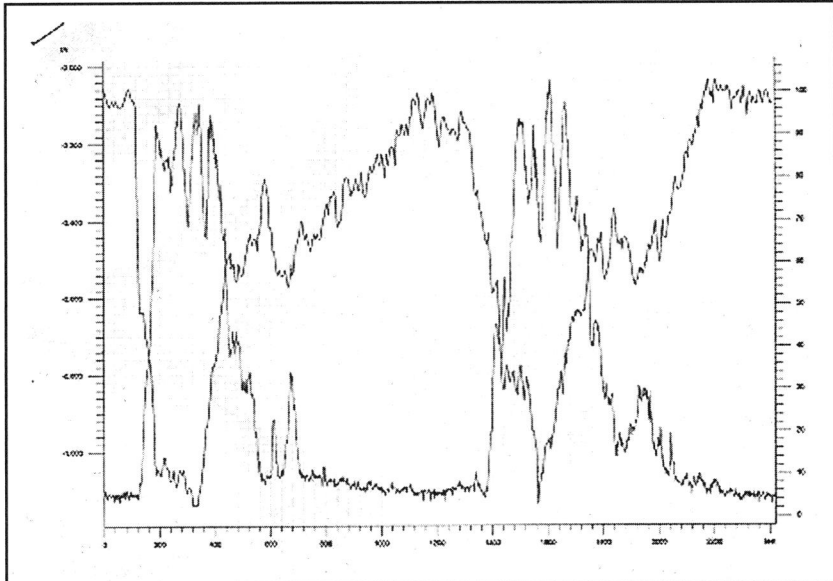

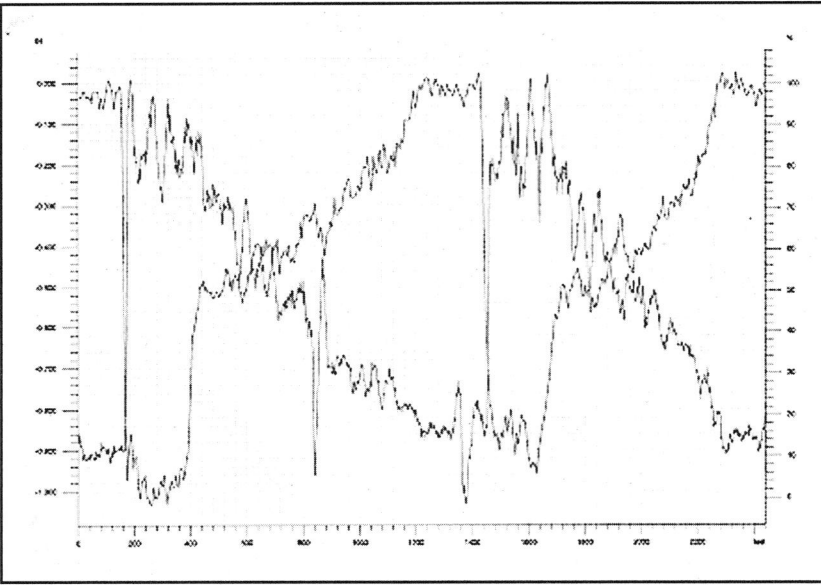

Different-length coil springs of the same spring rate were tested in the same car to prove or disprove that they make a difference in handling.

These charts of the rear-end movement in relation to the throttle position were made with data gathered from the PI system. The chart on the top is with the old pull bar and shows how the pull bar bottoms out quickly. The chart on the bottom is with the stiffer pull bar and shows a relatively smooth transition as the throttle is applied, which allows for more throttle application.

PI data-acquisition system, with the shock travel sensors. The softer front spring slows the rate of weight transfer to that corner, but the static weight on that corner depends on the ride height setting for the spring.

Many racers will run extremely soft springs in the rear of the car in an effort to get more forward bite. But, a soft right-rear spring can sometimes unload the left-front under acceleration. Also, the super-soft rear springs unload easily under deceleration, which can make a car push on corner entry. Often,

a racer will use a tie-down shock on the left-rear corner in an effort to minimize the push. All this usually does is lift the left-rear tire into the air upon entering the corner. Now, the traction-limited front end, which was pushing, doesn't feel as bad because the rear end is now traction limited. Great, you've made your car slower, but at least it feels better!

So, these changes were made and times were improved. In fact, one car that received no tuning ran the same lap times all day, and by the end, the cars getting worked on were running almost four tenths of a second faster. That's a lifetime in the Late Model ranks.

SPRING-LENGTH TESTING

An old track tale that is constantly being recycled is that the handling of a dirt car can be tuned if you switch from a long-coil spring to a short-coil spring (or vice versa) with the same spring rate. To test this, Weaver's car, which had been optimized with the use of the data from the PI Research system, was equipped with a 12-inch spring and a run was made. Then, a 10-inch spring was installed and the car taken back out on the track. No noticeable difference was seen in performance or by the driver. What the Hyperco people feel racers are experiencing is coil bind of the short spring. The coil bind would increase the spring rate by reducing the number of usable coils and would dramatically change the suspension setup. This is more of a usage problem than a spring problem. The lesson here is that if you are switching between different-length springs, make sure you don't require more travel than the spring can provide.

SEARCHING FOR FORWARD BITE

Forward bite, the traction the rear tires have when the car is accelerating off the corner, is as important as any other dynamic aspect of the race car on the track. How to increase forward bite is not clearly understood. One of the most common ways to attempt to increase forward bite is to reduce the "shocking" effect to the tires caused by the sudden application of power. This is done by allowing the rear end to rotate in a controlled manner. The most common way this is done in the dirt Late Model ranks is with a component called a pull bar. The rear suspension is designed to allow the rear end to rotate freely, but the pull bar is attached between the frame and a bracket bolted to the rear-end housing. As the driver applies the throttle, the rear end rotates, trying to extend the pull bar. Through track testing and swapping components, it was determined a pull bar with more resistance to rotation was needed. The final pull bar never bottomed out and the car achieved a 3-mph-higher top speed!

Hopefully this information should get you thinking in the right direction. If it means anything, you should know both Weaver and Moyer won their next races using the information gained at this test. Let's hope you can win with it, too.

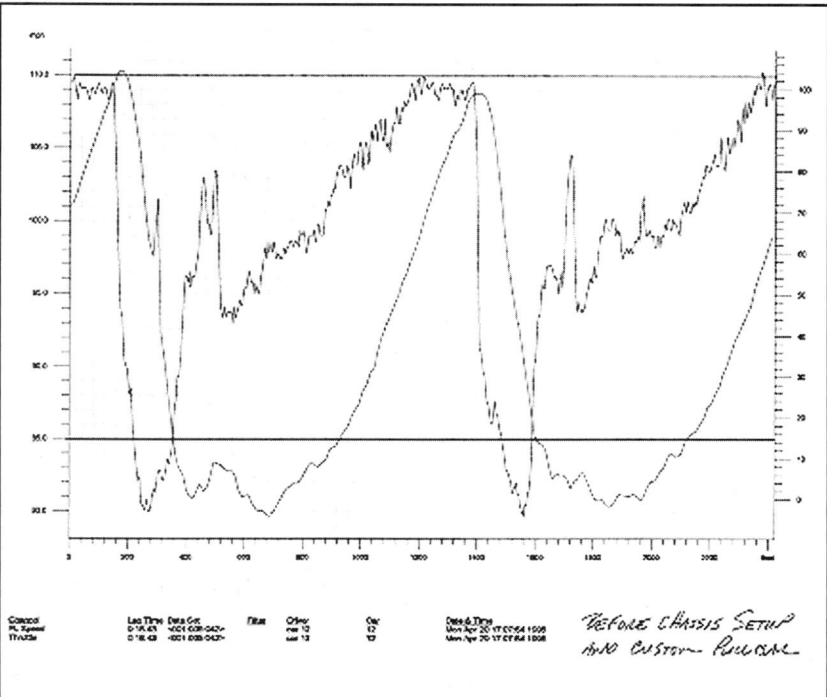

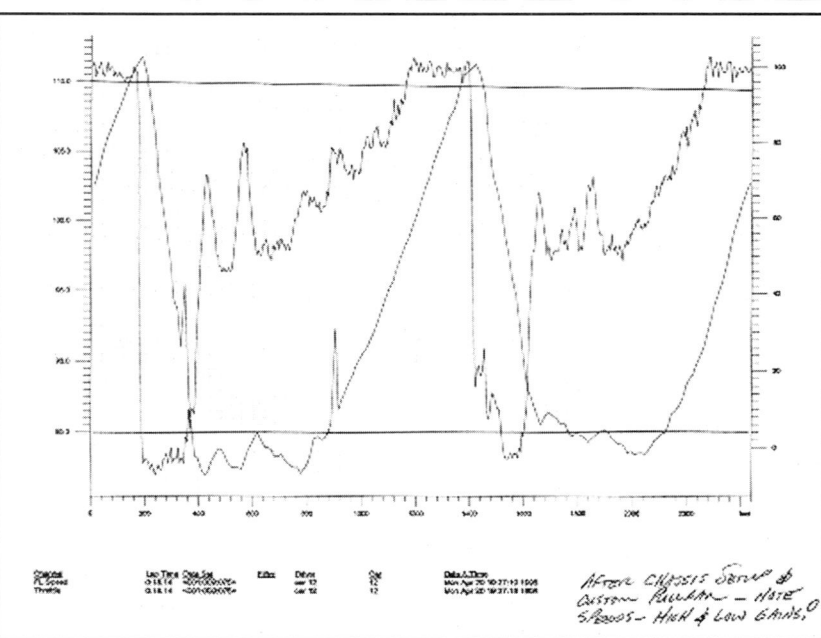

These charts show the before-and-after data of the vehicle speed and throttle application from the beginning of the day to the end of the day. The car improved in speed and throttle application.

Chapter 15
LOW TO HIGH

Changing Setups From a Low-Banked Track to a High-Banked Track

by Jon Fitzsimmons

Racing on dirt presents unique challenges for a chassis tuner. So many variables come into play—the type of dirt, how the dirt changes over the course of a night, the banking of the track, and the ability of the driver can make for some interesting setup dilemmas. It's part science, part luck, and part experience. In addition, a lot of chassis mythology can get thrown in the mix.

For drivers racing at several dissimilar dirt tracks in an IMCA Modified or similar race car, getting the car "just right" for a particular track can be the difference between being a contender and driving the car straight into the wall. To stay out of the wall and drive fast, a few changes must usually take place when switching between different tracks. Adjusting the baseline chassis setup on high- and low-banked tracks can be a relatively simple process in which the springs are changed and the rear steer is adjusted.

But there's more to it than that. As with any car, having an accurate starting point with the chassis baseline boils down to the details. We will follow the process of checking the details (e.g. ball joints, air pressures, ride heights, toe, and so on), then move on to the real meat of the matter—examining which components to change and what realistic numbers are for one, common example.

For insight into these chassis tuning issues, we conferred with Scott Bartell, who drives an an IMCA Modified at a variety of dirt tracks in Southern California. When it comes to dirt (he also races on asphalt in another car) he can be found at Bakersfield's 1/3-mile, high-banked clay oval, or Ventura's 1/5-mile, semi-banked dirt oval. At these two tracks, he's learned a thing or two about changing a setup to accommodate banking. He has also attended AFCO's chassis-tuning seminar.

We picked his brain to learn the difference between baseline setups on two different types of dirt tracks: high-banked and low-banked.

Scott Bartel (right) and Crew Chief Craig Bell (left) keep the #42 Modified dialed-in and ready to race. Bartel's chassis is a Dirt Works–made '68–'72 Chevelle with a monoleaf rear suspension and coil springs up front. The race car is going through the relatively easy process of being switched from a low-banked setup to a high-banked setup. Please note that all numbers mentioned here apply only to Bartels' car and are intended to be used as an example. This is not a guide to any other specific race car.

LOW-BANKED SETUP

This race car is going from a low-banked setup to a high-banked setup. Not much will change except for the spring rate and the rear steer. WARNING: The numbers used on Bartels' baseline settings (on a Dirt Works chassis) should not be applied to other similar cars. These numbers are provided as one example of what works for a single car. The untested results on any other car may not work at all. Please, don't configure your setup based on these numbers alone.

Left-front specifications:
Tire pressure: 9 psi
Spring rate: 650 lbs
Shock No.: 5/3
Frame height: 5 in

Right-front specifications:
Tire pressure: 15 psi
Spring rate: 850 lbs
Shock No.: 6
Frame height: 5 1/2 in

Left-rear specifications:
Tire pressure: 10 psi
Spring rate: 150 lbs
Shock No.: 4
Frame height: 5 in

Right-rear specifications:
Tire pressure: 10 psi
Spring rate: 150 lbs
Shock No.: 4
Frame height: 5 1/2 in
Rear steer: +3/8 in*

*Rear steer angle is related not only to the banking of the track, but also to the length of the track. As a rule of thumb, use less rear steer on longer tracks, and more rear steer on shorter tracks with less banking.

The first part of any baseline change begins with an examination of a few critical suspension parts. Check ball joints, measure ride heights, front-end alignment and scale the car to determine the exact starting point. Parts that are worn-out or not adjusted properly will cause the baseline measurements to be inconsistent. Start the process by checking the tire pressures.

Keep ball joints, chassis links and bushings lubed. This adds to the longevity of the parts, but it also keeps the suspension movement flexible and predictable. While doing this, check components for wear and tear.

Camber and caster settings need to be verified and adjusted, if necessary, to meet baseline specs.

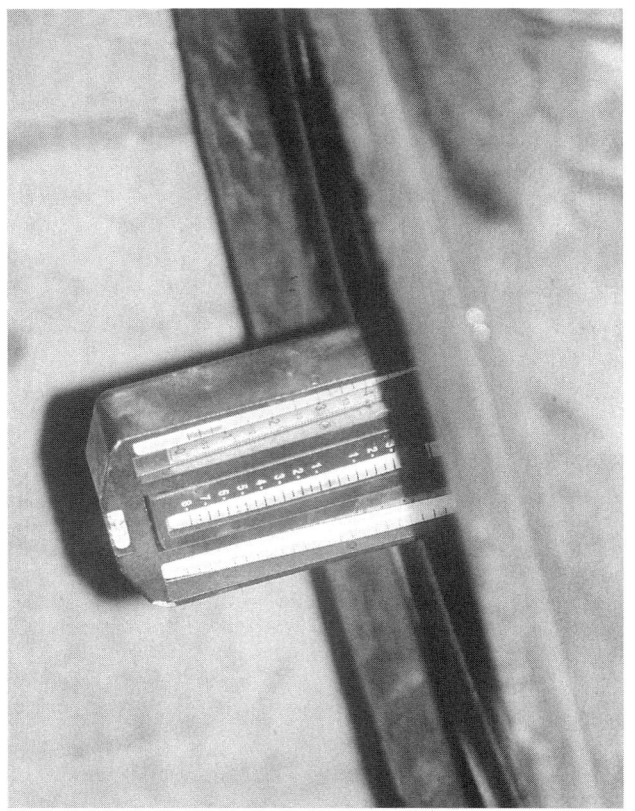

A quick and easy way to test a ball joint (to see if it is bent) is to hook up a drill to the ball joint's shaft and let it spin at a high speed. If it vibrates, then the ball joint is bent. A bad ball joint can drastically affect toe-in and toe-out settings. For the sake of a clear photo, we are showing a ball joint outside of the A-arm attached to the drill, but just spin the ball joint while it's partially mounted.

HIGH-BANKED SETUP

Left-front specifications:
Tire pressure: 9 psi
Spring rate: 650 lbs
Shock No.: 5/3
Frame height: 5 in

Right-front specifications:
Tire pressure: 15 psi
Spring rate: 850 lbs
Shock No.: 6
Frame height: 5 1/2 in

Left-rear specifications:
Tire pressure: 10 psi
Spring rate: 150 lbs
Shock No.: 5
Frame height: 5 in

Right-rear specifications:
Tire pressure: 10 psi
Spring rate: 175 lbs
Shock No.: 5
Frame height: 5 1/2 in
Rear steer: +1/8 in

The final settings didn't change much from the original. However, a few things need to be tweaked once the car is at the track. To increase or decrease the tightness and looseness of the handling, tire pressures are raised and lowered about a pound at a time. As a rule of thumb here, low pressures provide higher levels of grip than high-pressure tires; however, they also run hotter. Low-pressure tires tend to be more desirable on wet tracks. High-pressure tires tend to offer more benefits on dry tracks. In both setups, the camber and caster angles remained unchanged and at the chassis manufacturer's recommendations. The weight percentages also remained at the chassis manufacturer's recommendations. The key to chassis-tuning a dirt Modified for different tracks involves focusing on specific areas and not changing too many things around at once.

The first step to setting the car up for a high-banked track involves removing the 150-pound right-rear spring and replacing it with a 175-pound spring. Both AFCO No. 4 rear shocks were removed and replaced with No. 5 shocks.

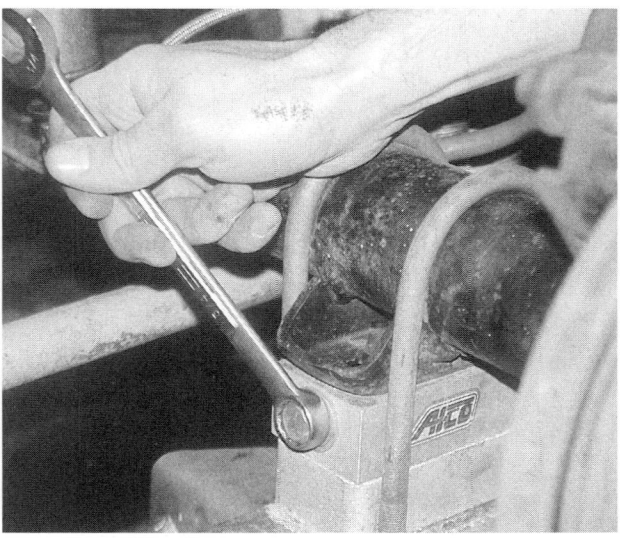

The rear steer angle was adjusted from the low-banked setting of +3/8 inch to the high-banked/longer track setting of +1/8 inch.

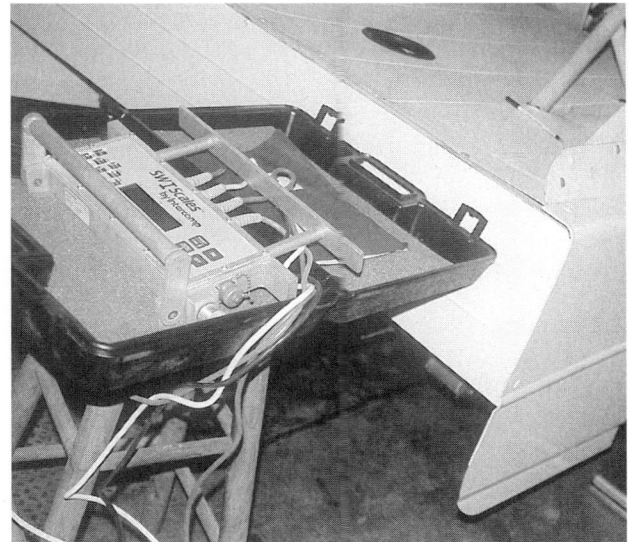

Once everything is put back together, the car is scaled to see where the different settings distributed the weight.

Adjustments are made to the weight jacks to bring the car back in line with baseline settings. Ride heights are double-checked, and the car should now handle better on higher-banked tracks.

Chapter 16
ON TRACK TUNING

Ten Dirt Suspension Tuning Tips

by Tom Rounds

Here is a typical four-link suspension on a dirt Late Model.

What is the best way to get your dirt Late Model to stick to the track and perform in a winning manner? We went to Keith Nosbish's race shop in Valrico, Florida, to find out. Nosbish races a dirt Late Model in the Southeast Motorsport series in Florida and at East Bay Raceway in Tampa. He comes from a family of racers, including his father, two brothers, and nephew. We also spoke with Mark Richards, manufacturer of Rocket Chassis, to learn some tips that will help every dirt Late Model racer.

"The track goes through different stages—from being heavy, loose dirt, to being packed with an asphalt-like feel as the races progress during the day," Nosbish explains. Because of this, there is a need for constant change to the race car's suspension to keep it on the track and be competitive with the other racers.

Here are several crucial tuning tips supplied by both Nosbish and Richards to help you get around the track faster than your opponent:

1. When you set up a modern dirt Late Model, it is best to start with the recommendations from the chassis builder, then adjust from there. This will assure positive results more quickly.

The Panhard bar or J-bar (arrow) can be adjusted in two ways. First, change the angle to increase or decrease the car's bite on the track. Second, adjust the overall height of the bar; this changes the roll center or the axis of the mass weight of the car. Roll center is changed for different weight distributions.

By moving the mounting location of the left rear shock, traction can increase or decrease. With the shock mounted on the back of the axle, the race car's traction will improve.

2. Analyze your handling problems correctly. You won't help matters by fixing a misdiagnosed problem, so think through your problems before you make changes.

3. Don't tune too tight. This can cause your race car to get too much traction and understeer. A common error with tuning too tight is that the driver may overcompensate in the turns, which gives the car a loose feel. This can result in a

CHASSIS ADJUSTMENT GUIDE FOR FOUR-LINK DIRT LATE MODEL

The following are setup suggestions. You may need to get a feel depending on your car and the track as to which of the available changes will be most effective.

	TIGHT ON ENTRY	LOOSE ON ENTRY	TIGHT IN MIDDLE	LOOSE IN MIDDLE	TIGHT ON EXIT	LOOSE ON EXIT
LF Spring	Soften	Stiffen	-	-	-	-
RF Spring	Stiffen	Soften	-	-	Soften	Stiffen
LR Spring	Stiffen	Soften	-	-	Soften	Stiffen
RR Spring	Soften	Stiffen	-	-	Stiffen	Soften
Panhard Bar @ Pinion	Raise	Lower	Raise	Lower	-	-
Panhard Bar @ Frame	Lower	Raise	Lower	Raise	-	-
LR Top Rod—Frame	-	-	Lower	Raise	Lower	Raise
LR Bottom Rod	Raise	Lower	Lower	Raise	Lower	Raise
RR Top Rod	Lower	Raise	-	-	Raise	Lower
RR Bottom Rod	Raise	Lower	Raise	Lower	Raise	Lower
Ballast Height	Lower	Raise	Lower	Raise	Lower	Raise
Left Side %	Increase	Decrease	Increase	Decrease	Decrease	Increase
Rear %	Decrease	Increase	Increase	Decrease	Decrease	Increase
Diagonal Wedge	Increase	Decrease	Decrease	Increase	Decrease	Increase
Trail RR Wheelbase	Lengthen Right Side	Shorten Right Side	Lengthen Right Side	Shorten Right Side	Lengthen Right Side	Shorten Right Side
Wrapup Shock	Lower @ Frame	Raise @ Frame	-	-	-	-
Air Pressure	Increase RR	Decrease RR	Increase RR	Decrease RR	Increase RR	Decrease RR

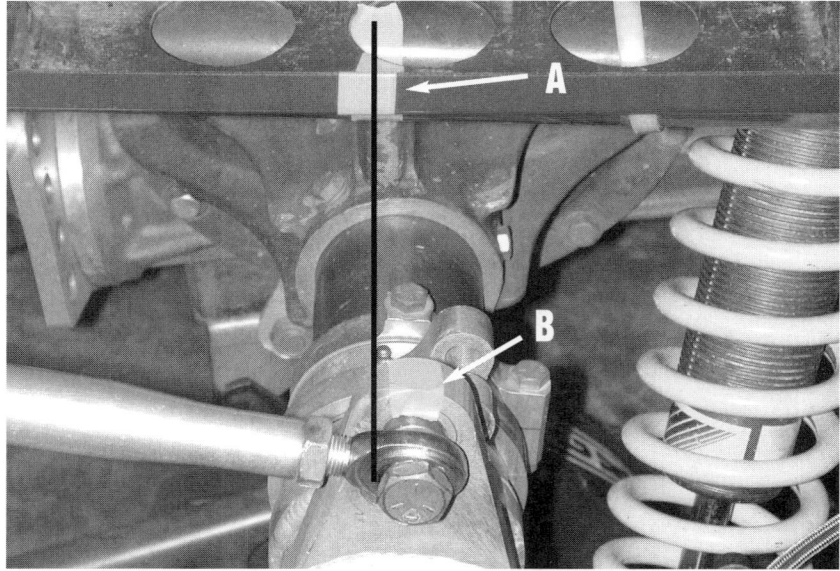

misdiagnosis and an incorrect adjustment.

4. Make sure you factor in tire choices when making adjustments. Proper tire compound, construction, and air pressure is a critical part of handling.

5. Be careful not to overadjust any suspension component; this can produce an effect that is the opposite of what is desired.

6. Make changes in one area at a time, so you can understand what you did and how it affected the race car. Making multiple changes is a shot-in-the-dark approach that usually nets poor results.

7. Shock tuning has become extremely important to help make a race car handle properly. Double-adjustable shocks make it easier to accomplish this part of tuning.

8. When running at a stop-and-go–type of dirt track, a tighter suspension is usually best.

9. When running a big-momentum type of dirt track, a looser or freer type of setup will usually help you run faster.

10. A database with accurate records of all your race car's changes and results will help eliminate continual trial and error. Include track information for better results.

Chapter 17

SETTING UP YOUR DIRT CHASSIS

The Winning Technology of Pro Dirt Teams

by Bob Bolles

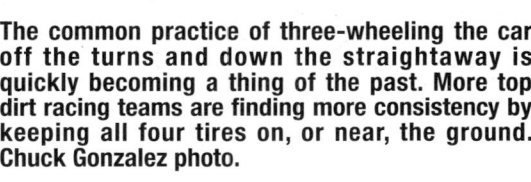

The common practice of three-wheeling the car off the turns and down the straightaway is quickly becoming a thing of the past. More top dirt racing teams are finding more consistency by keeping all four tires on, or near, the ground. Chuck Gonzalez photo.

Dirt car setup is far more complicated than setting up an asphalt Stock car. This book is all about dirt car setup and nothing else. It is so specialized and the techniques have developed so much over the past five years or so that it is time to spill the beans on what is working and why.

Setting up a dirt car is complicated, not only by the fact that the surface is difficult to work with, but mostly by the way it is constantly changing. You must be willing to make rapid changes to meet the requirements of the racing surface. Too many teams, surprisingly on the professional side of dirt racing, will come to an event and make few chassis adjustments when the track conditions continue to change drastically from practice, through qualifying, during the heat races, and finally in the main event. It's no wonder that the pole car seldom wins.

The routine for setting up a dirt car should start in the shop. A team must anticipate the conditions it will encounter at the next track it plans to race. Decisions concerning spring changes, tire selection and grooving, rear suspension adjustment, weight distribution, and even the front geometry should be finalized, and notes should be handy so the team can make deliberate and quick changes as they become necessary.

That's easy to say, I know, but it can and is done by many of the winning teams. It's easy to write one thing and allude to changes in technology as being "cutting edge" or "state-of-the-art," but what really governs the significance of any new technology is hearing it from the actual racers, especially the ones who often win. The ones we spoke with were in agreement that times have definitely changed.

"Racers these days are more technology minded," says Billy Moyer, one of the top dirt Late Model drivers over the past 20 years. "They want to know more about how their cars work. The joy of dirt racing is being able to learn how to make all of the adjustments work together to improve the car's performance." Moyer doesn't pretend to know it all and says each year is a learning experience for him. That can be said for us all.

Brian Birkhofer, runner-up in the competitive Xtreme Dirt Car Series in 2003, states, "There is a changing of the guard happening at the end of this year [2003]. What would have worked in the past is no longer good enough." Birkhofer says his team is working to make the car more balanced, learning the exact location of the moment center design. They have already experimented with different spindles and other adjustments to find the best configuration.

Let's take a look at the different elements to work with on a dirt car. For the most part, we'll present information about the Late Model designs, but much information can be utilized by all dirt racers, from Stocks to Modifieds.

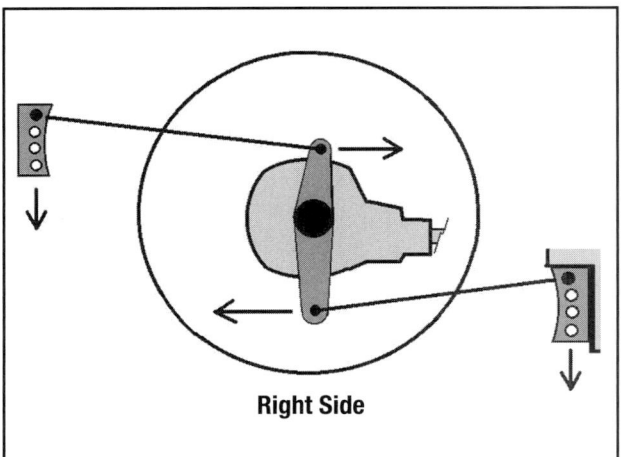

Right Side

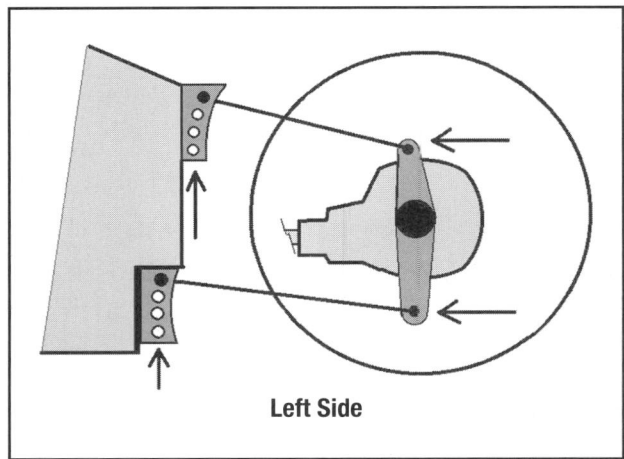

Left Side

The two most common types of rear suspension links are the Z-link (left) and the four-bar (right). Some teams will use a Z-link on the right side of the car and a four-link on the left side. This helps them control the rear steer characteristics between the different setups.

Most Late Model cars utilize birdcage-type attachment points where the linkages connect to the rear axle tubes. This allows the rear end to be free to rotate. There are two devices that can be used to control and regulate the rotation of the rear end as the car is accelerating and braking. They are the lift arm (shown on the left) and the pull bar.

ELEMENTS OF ADJUSTMENT
Front End Geometry

A good setup starts up front with the geometry of the front suspension. Don't begin to think this is not important, because it is. Furthermore, some of the top car builders in many different classes of dirt car racing, including the Modifieds and Late Models, have redesigned their front ends for better moment center location and camber change characteristics. Partial proof came to me as I was putting this story together. A team sent me the geometry data from its Late Model car purchased from a recognizable car builder; the car was very well designed. Only a couple of years ago, many other brands of this car type were terrible in their designs because none of us knew any better.

Know as much as you can about your front end geometry, and don't be afraid to change things to make it better. The most influential factor concerning the front geometry is the track's banking angle. The higher the banking, the farther the moment center can be located to the right in the car. MCs located to the left are useful for the flatter, slicker tracks.

Rear Geometry

There are probably several hundred different configurations we can use in typical four-link or Z-link rear suspensions. If we combine the different bar angle possibilities with the various spring placements, indexing properties of the birdcages, pull bar angles and spring rates, lift arm rates that regulate lift on acceleration and compression during braking, shock rates for the four corners, plus the fifth and sixth coil, we begin to get the picture. The adjustment combinations are even much more than those for a Formula One or Indy car.

We need to decide what we want to achieve and then set the bars, etc., so that the car will do what we want. The various linkage designs from the four-bar to the Z-link offer possibilities for changing the rear steer characteristics as well as the distribution of weight while decelerating and while under acceleration loads.

Each side can be adjusted with respect to "front to rear movement" as the chassis moves vertically, or to steer either direction depending upon what the chassis is doing on a particular side of the car. The four-bar can be set up to move the wheel

This IMCA dirt Modified utilized a Z-link rear suspension. Note there are two holes for the coilover mount on the lower arm. Moving the mount changes the motion ratio of the arm and the spring rate that is felt by the chassis. Using a 200–pounds per inch (ppi) rated spring, the front hole would yield a net spring rate of about 110 pounds. Moving the spring to the back hole would increase the net spring rate by 20 pounds to about 130 ppi.

The lift arm attaches to the rear end housing and runs forward to an attachment point. The movement of the arm is controlled by a spring and shock on acceleration and on a spring when the car is braking. The fifth coil, which controls the acceleration forces, is to the rear. The sixth coil, which controls the braking forces, is to the front.

considerable distances fore and aft, and that movement can either help performance or put you out of contention.

Rear geometry deals not only with wheel movement, but also the forces associated with acceleration and braking. As we accelerate, there is a great amount of torque applied to the rear axle housing, and dirt car manufacturers have designed several systems to take advantage of that force. Among the most popular systems are the pull bar and the lift arm. Both help cushion the torque related to instant application of power to help the rear tires maintain grip with the racing surface, and to control the forces of engine and wheel braking when decelerating.

Both systems can utilize a shock that will control the speed of movement both ways. This is essential, because a spring will react relatively slowly when being compressed, and violently when released from compression. The shock controls both, but obviously must do much more work in the rebound mode.

Weight Distribution

The changing of the static distribution of weight in a dirt car can be accomplished in two ways: by moving weight around in the car or by adjusting the distribution of weight on the four wheels without moving actual weight. The latter involves changing the height of the spring by turning the

adjusting ring at the top or bottom of each spring to regulate the amount of crossweight, or, in more familiar terms to dirt racers, the amount of left-rear (LR) weight.

LR weight refers to the number of pounds of weight supported by the left-rear wheel versus the weight supported by the right-rear (RR) wheel. One hundred pounds of LR would indicate that if the car were weighed, the LR scale would read 100 pounds more than the RR scale. Setups can be developed around a particular LR weight number or the LR weight around a particular setup to tighten or loosen a car.

In some cases, high amounts of LR weight can produce less traction than smaller amounts. We generally relate high crossweight, or LR weight, with improved traction, but that's not always the case. If we have a car that utilizes all four tires (all on the ground with sufficient weight on them to make them work), we can definitely use a higher amount of LR weight to try to balance the weight across the rear tires. The more equally loaded a pair of tires is, the more available traction.

This does not work in the case of a setup that lifts the LF tire off the ground. Once the LF is elevated, all of the front weight of the car is supported by the RF tire. Since diagonal tires share the loads in

For many top racers, the only time you'll see the LF tire off the ground is when the car is on jackstands. There are advantages to keeping the front of the car low into and off of the turns.

combination, if the RF weight goes up, so does the LR weight. If there is already a high amount of LR weight supported by that tire, imagine what it is after the LF comes off the ground. For this reason, we need to run a much lower amount of LR weight with a three-wheel setup.

Tire Grooving and Siping

As track conditions change, so do our needs pertaining to tires. We usually have the opportunity to change the grooving in the tires to increase the grip and to help control heat buildup. This is a highly complex subject and best left to experts such as Professor C. P. Furney Jr., who took the time and effort to help us dirt racers understand more about dirt Late Model tires by writing a book about it. *Selection and Application of Late Model Dirt Racing Tires* is a comprehensive and detailed discussion about the properties of dirt tires and how to utilize them for different conditions.

We can change the characteristics of our dirt tires by understanding how the conditions of the track, at any given moment, affect the surface rubber on the tire. The idea is to help the tire produce the maximum allowable grip by making appropriate changes to the tread pattern by cutting (siping) or grooving. How do you do that? Read the book.

Dirt Car Aerodynamics

Most dirt racers probably do not agree that aero has a profound effect on dirt cars. For the most part, they are more right than wrong. On the other hand, aero may play a major role in dirt car performance, and our understanding of it may answer some nagging questions as to how certain setups work well and others do not.

When we examine the shape of the dirt car, we see large, flat sides, wedge-shaped noses, large rear spoilers, and open rears around the bumper. These

are shapes that can produce considerable aero effects, both good and bad.

If you have ever observed a dirt Late Model car cruising through the pits at around 10 or 15 mph, you may have noticed a lot of dust billowing out of the back of the car. Have you ever wondered why there is so much disturbance of the dust on the ground? A pickup truck can go by and not move any dust, so what causes the difference? It is the suction created by air being pulled out of the back of the car as it moves through the air and sucks the dust up off the ground—and this is at only 10–15 mph. Think of the effect at 60–80 mph in the turns.

So if we agree that there is a low pressure effect and with it some amount of downforce, we might try to utilize that effect to our advantage. The car's attitude works to improve downforce or destroy it. Many racers struggle with a car that pushes on exit when the LF wheel, and the nose, is hiked up a foot or more off the ground. No wonder it pushes; the car is catching a lot of air under the LF, and that might cause aero lift instead of aero downforce.

The car might stick better up front, through the middle, and off the corner if we can keep the LF corner down and allow the aerodynamic downforce to work. All is going to be lost once that LF comes up as we accelerate.

ADJUSTING THE SETUP FOR TRACK CONDITIONS

Tight, Moist

TracksMost dirt tracks start an event with a lot of moisture due to the crew watering and/or tearing up the surface prior to the event. The conditions do not usually change much from practice to qualifying, and the surface may become even more moist if the crew waters the track prior to the qualifying runs.

The trend among top racers has been to run more even spring rates across the front and rear and a more level track bar to balance the setup when there is a lot of traction. The bar may even be mounted on the right side of the chassis for more consistency. We need all of this when the tracks more closely resemble asphalt conditions as the G-forces increase. We can then utilize all four tires and a more level body configuration relative to the track surface to improve the aero downforce that exists.

The rear geometry needs to be arranged so that minimal rear steer takes place in order to keep the car going straight ahead, much like an asphalt car. With all four tires on the ground, a high amount of LR weight can be put in the car to provide improved traction off the corners. The car will turn in well and drive through the middle because of the balanced setup causing the LF tire to work to turn the car. Some racers have even been known to run a stiffer RR spring when conditions warrant less RR chassis travel.

Mid-Point Traction

As the track begins to dry out through the qualifying heat races, a team must observe the surface conditions during the heats just before they race. If the track is going black-slick, the setup may need to change as well as the tire selection. These slicker track conditions can cause excess heat buildup in the tires, and more siping is required to help cool the rubber.

The RR can be a little softer to promote traction off the corners, but that will produce more chassis roll at the rear and unload the LF tire. The LF is not going to hike up, but it will become less loaded. So the crossweight, or LR weight, will need to come down so that the balance of weight between the rear tires doesn't become uneven. This prevents a condition in which the LR (as well as the RF) ends up with a high weight while the RR supports very little weight. Less equally loaded rear tires mean less traction off the corners.

Dry, Slick Conditions

As the track dries out further and the surface becomes dry and slick, drastic measures must be taken. We need more weight transfer, a sideways attitude of the car relative to the direction of travel, and a rear geometry that will help overcome the tendency of the car to push on entry and through the middle.

We would soften the right-side springs to help the front turn and the rear to promote traction off the corners. The track bar must be mounted on the left side of the chassis and angled to a greater degree (left side higher) to help pin the RR tire. The greater angle of the bar also causes the chassis to hike up in the rear, which produces rear steer, the degree dependent on the angle of the linkage bars in the rear suspension.

While all of this is happening, we need to make sure that the LF doesn't lift too high off the ground. Recall the aero discussion about keeping the nose close to the ground to keep the downforce in effect. The LF tire can be floating somewhat, but if it is not well off the ground allowing air to get under the front end, the aero will still work.

A considerable amount of rear steer (steering to the right, usually utilizing the LR coming forward) may promote the aero effect, which helps keep the

The rear spoiler found on most dirt cars produces downforce at the rear of the car, which promotes bite off the corners. Elevating the rear of the car through whatever means, such as using a left-side, short, high-angle Panhard bar mount so the LR corner is jacked up in the turns helps the aero effect of the spoiler. This may be a positive as long as the LF does not come up as well to make the car push off the corners. For every positive effect, there is a limit as to how far we can go.

car down on the inside of the track in the corners. This comes from the flat sides of the car striking the air, similar to the effect of a wing on a Sprint Car. Once the car has moved through the middle and it is time to accelerate off the corner, we need the LR to squat and the tire to come back into a more straightforward position for better exit performance.

Most racers now agree that all rear steer should be accomplished through LR wheel movement. If the RR wheel moves excessively, the feel is inconsistent to the driver because the RR is normally the dominant driven wheel off the corners. The movement of the LR wheel is much less offensive in nature.

Conclusion

In a particular event, we may experience a wide variety of track surface conditions that require us to adjust our setup. A team that makes an educated guess about the changes to make has a better chance of winning than a team that makes very few changes. It takes a lot of work to make these changes, and we must think correctly in order to make the best changes.

In the long run, if we expect a fighting chance at winning, we must learn to react. After all, the top teams have definitely learned, they make the adjustments, and, best of all, they win often. Now, let's detail how to make those adjustments.

We should never be afraid to make changes to our setups in order to match the changing track conditions that are so common when racing on dirt. Today, the top teams are not only making changes, but have learned exactly which ones will make the car better as the track surface evolves.

ADJUSTMENTS FROM PRACTICE TO MAIN EVENT

Change, change, everywhere a change . . . a message from social protests of the '60s, could have been written in reference to dirt track racing. Throughout an event, most dirt tracks change surface characteristics, some in a dramatic way. Granted, there are some so-called dirt tracks that are always oiled and maintained to the same grip. These are exceptions and won't be addressed here. They are more like asphalt anyway, and a team racing a dirt car on those tracks would do well to set up their cars more closely to an asphalt car setup.

The most difficult tracks are the ones in which the type of dirt or clay needs a lot of moisture in order for the track crew to be able to groom the track. A promoter will often plow up a track during the week preceding the event, water it throughout the morning of the event, and then roll it while still pouring water on it until practice time. From then on, unless it is watered again, the track will begin to dry out and go through several phases of moisture content.

What makes these various conditions so hard on our setups is that the g-forces are changing along with the declining availability of grip as the track begins to dry out and go slick. As the g-forces change, our setups must change too. The difference is in the grip factor and mostly deals with the level of moisture present in the first inch or two of the surface material. Here is a list of phases of grip that a dirt team might encounter on a typical race day, along with the associated estimated g-forces that the car might experience:

1. Wet, sloppy, and rough: 1.6 to 1.7 g's (peak g's but very difficult for the driver to maintain a good line).
2. Very moist but graded and compacted: 1.5 to 1.6 g's (the best the track will be all day and very fast through the turns).
3. Hardened and more packed as the surface begins to dry out: 1.3 to 1.5 g's (still good corner speed while losing a little grip off the corner).
4. Black slick, a condition in which the track still has enough moisture to keep the material packed, but has hardened and is taking rubber: 1.1 to 1.3 g's (becoming difficult entering and turning in the middle of the turns).
5. Dry, a condition in which the top layer is now drying out and losing material in the form of sand: 0.9 to 1.1 g's (loss of grip is now getting substantial, and the car will not turn well or get bite off the corners. Driver touch is most important in maintaining momentum).
6. Dry slick, a condition in which the track has lost considerable moisture and is now very dry and slick

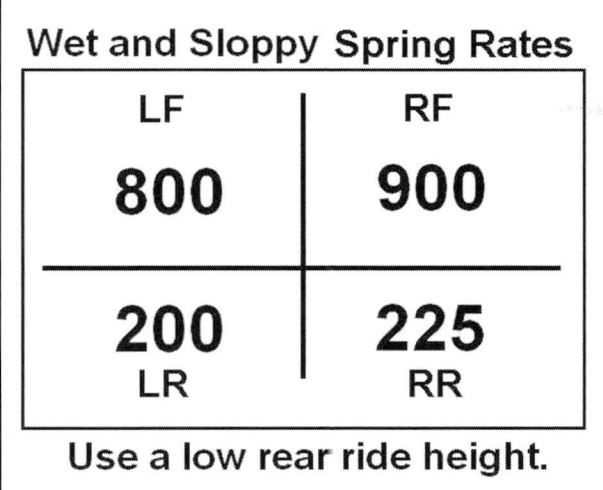

Wet and Sloppy Spring Rates

LF	RF
800	**900**
200 LR	**225** RR

Use a low rear ride height.

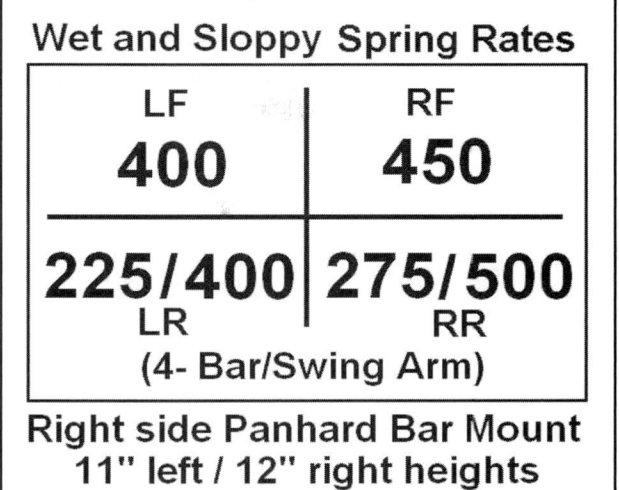

Wet and Sloppy Spring Rates

LF	RF
400	**450**
225/400 LR	**275/500** RR

(4- Bar/Swing Arm)

Right side Panhard Bar Mount
11" left / 12" right heights

The Hobby Stock setups (left) and the typical Late Model setups (right). For wet and sloppy conditions, usually the first situation we encounter during practice and possibly qualifying, we need to set up the car stiffer to withstand the high g-forces as the car gets thrown around in the ruts.

with little grip: 0.7 to 0.9 g's. (The cars are now very slow in getting into and through the turns, and traction under power is very low. Small gains in mechanical grip can provide substantial gains in lap times).

We can see how it is possible to experience a high 1.7 g's all the way down to a very low 0.7 g's in a single day. There is no way a team can run well in all of the conditions while maintaining the same setup. What we need to do is think about how these conditions, one by one, affect a car and then set it up accordingly. We will examine some physical rules that deal with chassis dynamics so we can arrange the setup to meet the conditions.

Let's use two of the most popular types of dirt car—a touring Late Model and a metric-style Hobby Stock car. For each type of car, we will set up for the six conditions listed above. The changes will be realistic and will not involve cutting or welding the chassis. Time is a factor and we want to make positive changes that can be accomplished in a short period of time.

Wet and Sloppy Setup

This track condition means the tires are getting a lot of grip and the surface tends to have ruts that grab the car hard. To be able to control the car, we need much stiffer springs and a more balanced setup front to rear. The car's center of gravity (CG) needs to be lower when the grip is high. A low CG number for a dirt car is around 17–18 inches off the ground.

Hobby Stock (HS)—We might run a stiffer right-front (RF) spring and a stiffer right-rear (RR) spring. A typical spring layout would be: left-front

(LF) = 800; RF = 900 (or a 750 with a full spring rubber, if spring changes are not allowed); left-rear (LR) = 200; RR = 225 (or a 175 with two full spring rubbers). The rear ride height would be set as low as possible, as this would raise the rear moment center and lower the CG to reduce the rear roll angle versus the front roll angle to help balance the setup. Any moveable weight would be mounted low and left in the car. If you have a choice, run tires with large grooves such as the all-terrain types for mud and snow. Run a higher pressure in all of the tires.

Late Model (LM)—A similar situation would apply to the LM car. For now, use a stiffer (RR) spring with a high Panhard bar mounted on the right side for consistency. Keep the weight low and to the left, and use tires with wide grooving to release the wet mud easily. Keep the air pressure high.

A typical setup might be: LF = 400; RF = 450; LR = 225 (400 pounds swing arm); RR = 275 (500 pounds swing arm); 56 percent left-side weight; Panhard bar at 11 inches left and 12 inches right, mounted to the right side of the chassis; and no front Ackermann with zero rear steer. On this car, we would have already moved the upper mounts for the rear shock/springs out as much as possible to reduce the rear spring angles.

Very Moist and Graded Setup

On a track that has good moisture and is smooth, the car will be easy to control going through the turns and the speeds will be high. We keep the low CG (mount the weight low in the car) to reduce weight transfer to the right side to help maintain

To maintain zero rear steer through chassis motion, we need to arrange the rear links at certain angles. The swing arm Z-link (left) as well as the four-link setup (right) are set so that the birdcage ends at the top and bottom will move in opposite directions as the chassis compresses on the right side and rebounds on the left side. This keeps the axle tube in the same position front to rear to provide zero rear steer.

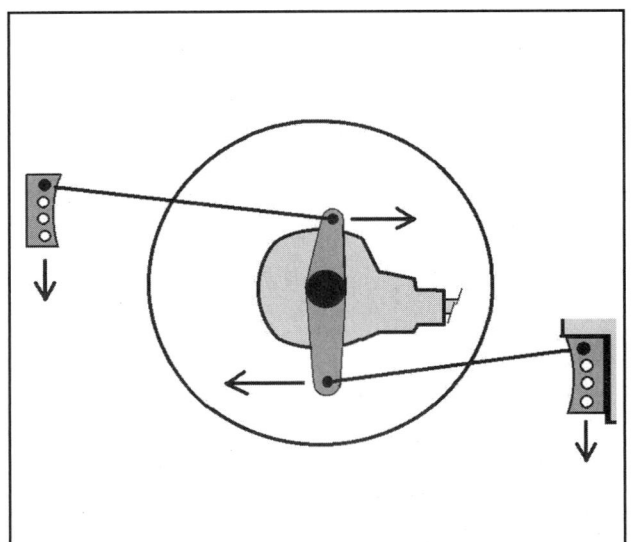

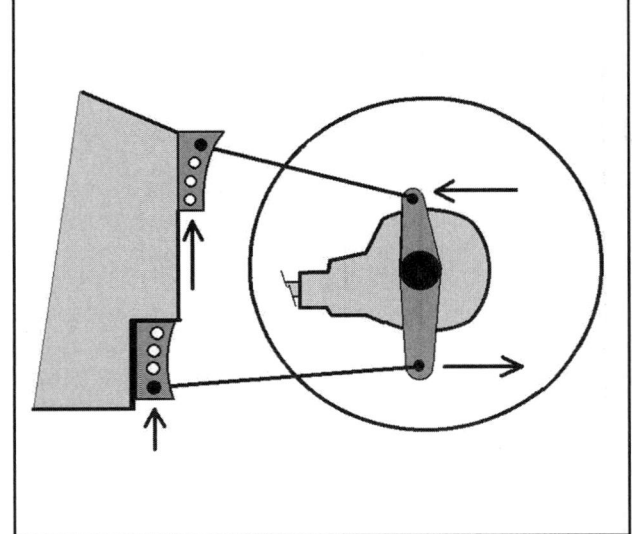

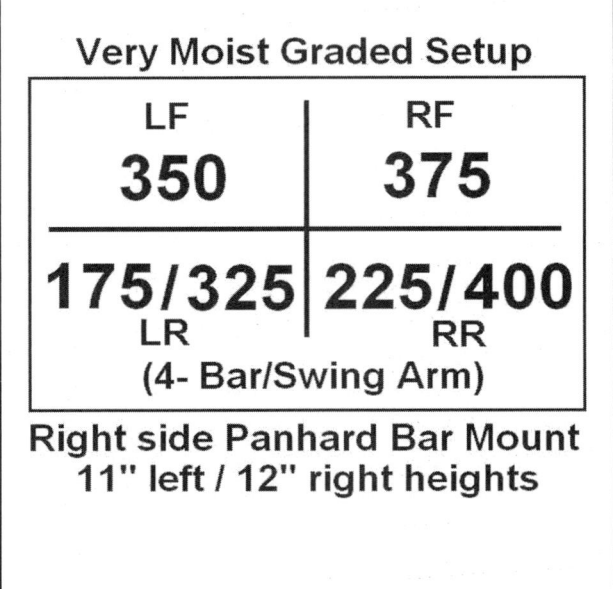

Very Moist Graded Setup

LF	RF
350	**375**
175/325 LR	**225/400** RR

(4- Bar/Swing Arm)

**Right side Panhard Bar Mount
11" left / 12" right heights**

The Late Model setup is very similar to an asphalt setup when the dirt has a lot of traction. The car can be driven straight ahead, deep into the corners, and hard off the corner. The right-side chassis Panhard mount technique makes the car more stable for the high g-force conditions. We will move the bar over to the left later on.

left-side weight in the turns.

HS: We could maintain the previous setup as the track begins to smooth out, but it will keep a considerable amount of grip. We could raise the rear ride height an inch or so to be ready if the track begins to dry out some and affect bite off the corner. Keep the all-terrain tires and lower the pressures a bit.

LM: These conditions are the fastest you will encounter. The car should enter, go through the middle, and come off the corners going straight ahead and balanced in the overall setup. We will make some changes to the previous spring setup. A typical setup might be: LF = 350; RF = 375; LR = 175 (325 pounds swing arm), RR = 225 (400 swing arm); 56 percent left-side weight; Panhard bar at 11 inches left and 12 inches right; and no front Ackermann with zero rear steer.

We will need to change to tires that are harder, with less gap and more surface rubber. The air pressure must come down some, but do not run soft on air at this point. Zero rear steer is essential now because our dirt car will handle much like an asphalt car. The driver can drive straight into the corner, brake harder, and get back into the throttle sooner than at any other time of the day.

Under these conditions, teams using softer RR springs with Panhard bars high on the left side and low on the right side and with plenty of rear steer will find trouble. In short, they will be all over the track.

Hardened Track Setup

Once the track has hardened, it will quickly continue to change toward black and then to slick. The hardened track condition may only occur for one heat or so. We might want to anticipate the dry and slicker conditions coming now for lack of time later.

HS: As the track begins to dry out, the surface will start to become black and a little slick. We can now begin to soften the setup somewhat and change the RF spring to an 800-pound (rated) spring or remove the RF full spring rubber and replace it with a half rubber. The RR spring should change to a 200-pound (rated) spring or pull one rubber out. Raise the rear ride height another inch.

```
┌─────────────────────────────────┐
│      Black and Slick Setup      │
│  ┌───────────┬───────────────┐  │
│     LF      │      RF          │
│    375      │     350          │
│  ───────────┼───────────────   │
│  185/325    │  175/300         │
│    LR       │      RR          │
│   (4- Bar/Swing Arm)            │
│                                 │
│  Left side Panhard Bar Mount    │
│  12" left / 8" right heights    │
└─────────────────────────────────┘
```

Once the track has developed a black groove, we are fast on the way toward a slick track. The groove is usually narrow, and the area outside the groove becomes slick as the material dries out. The right-side springs are now softer than the left-side springs.

This does two things: It lowers the rear moment center and raises the car's CG. The tires can be changed to more of a street performance tread with slightly lower pressures.

Evening up the front springs will help the car turn, and softening the RR spring, lowering the moment center, and raising the CG will provide better bite as the track dries out.

LM: We can now begin to change the spring rates at the RF, RR, and LR. The RF = 350; RR = 200; LR = 200 (375 pounds swing arm). We can move the Panhard bar over to the left side of the chassis with the heights at 11 inches left and 9 inches right. We can start moving weight over to the right side to provide a lower left-side weight of 52–54 percent. Now softer tires can be used with some siping. Slightly increased Ackermann and small amounts of rear steer (LR tire coming forward) will help the car turn through the middle.

Black and Slick Setup

The track has now blackened over and the edges of the groove are loosening up with the surface material turning to drier sand. Passing becomes difficult because of the one-groove nature of the track. The "slide job" works well here because as you go under a car coming into the corner, you pass and inevitably slide in front of the passed car. Once your car hits the black groove, it will grip and you can accelerate off from the passed car.

HS: When the track turns black, the surface has become hard with very little loose material. The

moisture is evaporating off the first inch of material, and there will be less available grip without the surface becoming overly slick. We will still need to drive the car straight and try not to soften the right-side springs too much yet. We won't change the setup at this point in time, but the driver will need to think more about throttle control coming off the corners to prevent wheel spin.

LM: We will want to soften the setup a little more while the track is transitioning through this phase, and we need to reverse-split the front springs. We could continue to move weight up and to the right in the car to raise the CG about 1 inch and further decrease the left-side weight percentage. The LF spring = 375; RF = 350; LR spring = 185; and RR spring = 175.

The Panhard bar, mounted on the left side of the chassis, should be at heights of 12 inches left and 8 inches right. Tire selection should include a little more siping to help dissipate heat, as the tires will begin to spin slightly coming off the corners. There will be less grip, but not to the point of going overly soft on the springs or introducing great amounts of rear steer to turn the car.

Dry Conditions Setup

The track has now dried up and the black is mostly gone. The surface has not come apart just yet, and passing is a bit more manageable because the entire width of the track is uniform in grip. If the track changes banking with more angle on the high side, then passing can be done on the outside more easily than down low if the car is handling well.

HS: We can now soften the right-side springs, or pull out all of the spring rubbers. It is time to soften the car, move weight over to the right, and lower the rear moment center more. Change the RF spring to a 750-pound spring and the RR to a 175-pound spring. Raise the rear ride height one more inch, move weight up and to the right in the car, and soften all of the shocks. Reduce the air pressure in the tires by 2 pounds all around.

LM: It is time to go a little more radical in the setup. We will soften the RR spring and the front springs. LF = 350; RF = 325; LR = 200 (350 pounds swing arm); RR = 175 (325 pounds swing arm). Raise the Panhard bar by 1 inch to 13 inches left and 9 inches right. Increase the rear steer by changing the angles of the left trailing arm. Decrease the air pressure all around for more footprint area, and groove the tires for the most side bite with more grooves and a limited number of siping cuts.

Now the track has gone to an overall slick condition that requires more imagination. We can utilize the entire width of the track for passing or to find the fastest groove. Bite off the corners will be at a premium.

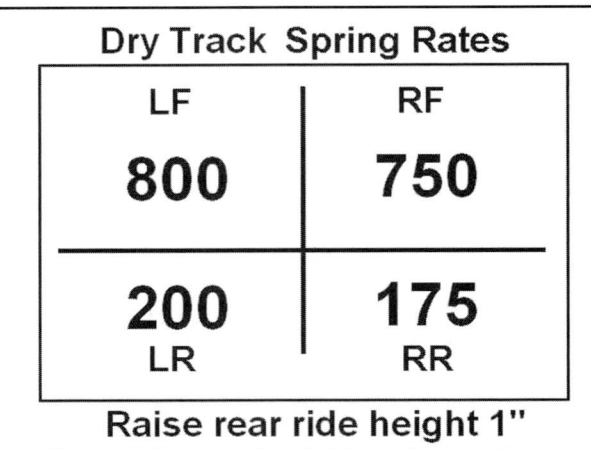

Dry Track Spring Rates	
LF	RF
800	**750**
LR	RR
200	**175**

Raise rear ride height 1"

Dry Track Setup	
LF	RF
350	**325**
200/350	**175/325**
LR	RR

(4- Bar/Swing Arm)

Left side Panhard Bar Mount
13" left / 9" right heights

We have now reached the final stage in the track's transition. The surface has dried out completely, and fine particles of material (sand) cover the surface, reducing the traction to very low levels. We need to incorporate rear steer in greater amounts to improve entry performance as well as take advantage of the aero affect of running the car at an angle (top view) to the oncoming air.

Dry Slick Spring Rates	
LF	RF
750	**650**
LR	RR
200	**150**

Dry Slick Track Setup	
LF	RF
350	**325**
185/325	**150/275**
LR	RR

(4- Bar/Swing Arm)

Left side Panhard Bar Mount
12" left / 8" right heights

Install softer rebound left-side shocks and softer compression right-side shocks. Raise the moveable weight up and to the right side as much as possible for a 50–50 percent side-to-side weight distribution. Increase the Ackermann to 1 degree in 15 degrees of steering. The LF tire will be doing less work with these changes but should still be touching the ground, and the Ackermann will help it provide turning power.

Dry-Slick Setup

Now we have come all the way to the most difficult conditions we will see in dirt track racing. We must get radical now, but not to the point of jacking the LF corner of the car off the ground. We need to maintain the downforce our car produces for corner speed, while providing some kind of bite off the corners.

HS: With the track going to full dry and slick conditions, we need to get more radical in our setup. We will now soften the LF spring to a 750, the RF spring to a 650, the RR spring to a 150, and move all of the moveable weight to the right side and higher up. We will take more air out of the

tires, but not enough to cause them to roll over. We should soften the shock rates all around so that the "shock" to the tires will be reduced as the car transitions going into the turns and accelerates off the corner. Smooth driving is important now as any complete loss of grip in the front or rear tires will cause a great increase in the lap times. Soft driving seems slow, but the alternative is always slower.

LM: We definitely need to soften up the setup and introduce more rear steer to be able to negotiate the turns without causing the LF tire to be lifted completely off the track surface. When the LF is hiked up considerably, we lose any front downforce that we might have enjoyed, as a great deal of air comes under the raised LF spoiler and we lose most of our front grip. There is only one way to stop a severe pushy or loose car, and that is to lift off the throttle. We can observe cars that three-wheel off the turns and hear the throttle being cycled on and off.

The setup may look like this: LF = 350; RF = 325; LR = 185; RR = 150 (double the rear springs for the swing arm car); Panhard bar at 12 inches left.

A more sideways driving style is sometimes helpful for mostly aerodynamic reasons. Putting the large, flat side of a Late Model into the wind has a pronounced effect on slowing the car and creating lateral lift. Lower pressure on the left side will provide a lateral force that counters the centrifugal force that is trying to take the car up to the wall. That is probably why rear steer (left tire forward) is sometimes so helpful. As the car rotates, it develops an aero force that acts like added traction. This effect can be taken to an extreme, and we never want to three-wheel the car in order to create the effect.

In our journey through a full day of dirt track racing at "Anywhere Speedway," we have purposely left out a number of setup variables such as fifth and sixth coil adjustments, gear changes, birdcage indexing, rear spring placement (front or rear of the axle tube), the decision to clamp or not to clamp, and others. Opinions will vary on these topics. Our intent was to provide some thoughts on making changes to the basic setup that might improve the car and make each phase better.

The key, again, is to be sure to make changes based on your track conditions related to your type of car. To run a single setup throughout the day is not going to get the job done in this day and age. Racers of all classes are getting smarter by the year, and getting to the front is a matter of staying on top of the most current technology. It's what racing is all about and it's all fun. If we had nothing to learn, we may as well go golfing—not.

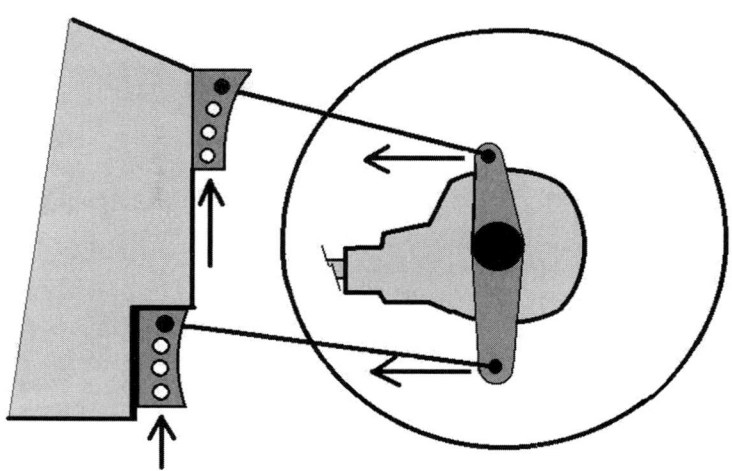

We can introduce a lot of rear steer, if needed, by positioning the left-side links in the four-bar car all the way to the top of the adjusting holes. We usually do not want the RR to steer because that is the driven wheel off the corners and needs to remain stable in its position fore and aft.

Chapter 18

DIRT TIRE GROOVING

by Scott Bloomquist
Photos and illustrations by Tom Hintz

On this night, Scott watched the other cars and the track until just before the feature prior to deciding on his grooving pattern. He won the race by more than a straightaway.

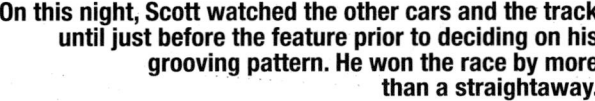

The critical point every racer has to remember is that the tires are the only link between everything you've done to prepare the car and the racing surface. Without a solid, predictable link between the chassis, engine, and track, an otherwise perfectly set up race car cannot perform to its potential. By using grooving techniques appropriate for dirt track conditions, racers can tune their tires just as they tune the chassis or engine to get the most from their race car.

Much like an asphalt racer, dirt racers have to pick the right tire compound. But there is another important step to getting the best performance from a tire on dirt tracks that can be expected to change drastically from week to week, or even race to race. Dirt racers often have to alter the design of a tread pattern to best work with the conditions at a particular track on a particular day and with their driver's style or habits. We can't tell you exactly what specific kind of grooving will work best for your situation, but we can provide basic information to make racers more able to decide for themselves as they encounter different racing conditions.

Some of the factors we have to think of are the abrasion of the track, how wet or dry the track is, and if the surface contains rocks or other debris that will tear up the tire. All of these conditions must be considered when deciding what

kind of groove to use, how deep to cut, and how many grooves can be used without causing the early demise of the tire. Along with these factors, racers must consider the amount of heat a track puts in their tires. In addition to enhancing traction, grooving helps the tire dissipate heat and can be used to help control tire temperatures.

SQUARE, "V" AND SIPE GROOVING

There are three basic shapes used in grooving: square, "V" and sipes. Square grooves are the same width through their entire depth, although most cutting blades actually have a rounded bottom. "V" grooves start out wide at the top and taper to nothing at their bottom. Sipes are thin slices made by installing the cutting blade upside-down in the holder and using the separate ends of the blade to cut thin slices in the tread. All three types of grooves can be used in various depths and numbers, depending on conditions and the length of the race.

Square grooves can also be used in various depths, but very wide grooves could provide enough leverage to tear the tread blocks if the track becomes sufficiently abrasive. Square grooves can be used on a variety of track conditions to produce sharp edges that bite into the track surface.

The "V" groove is often used when the track is expected to need more tread-contact surface later in the race. As the tread wears down, the grooves become smaller or disappear completely, depending on how deeply they were cut.

Siping is generally used to make the tread more pliable to better conform to irregular track surfaces. Siping does not produce the edges that increase traction that square or "V" grooves do. Siping also helps the tread maintain a more consistent wear that prevents glazing and helps keep the tire working uniformly.

GROOVING AND WEAR

Any time a groove or sipe is cut into a tire, it accelerates wear. The trick is to balance the benefits of grooving with the increased wear. One of the reasons a tire is siped is to prevent the tread surface from glazing over and becoming slick. Sipes keep the surface wearing and the tire working throughout the race.

Learning to predict the amount of wear you can expect from a track not only helps choose the right compound tire for the race, but it is important information for deciding what pattern, style, and depth of grooving to use. The object is to maintain the highest level of traction throughout the race without causing so much tread wear that the tire goes away five laps before the end of the event.

With an understanding of the basics of tire grooving, racers will be able to look back on their experience with tracks they have raced on and make better decisions on what type of grooving will help them most. Keeping records of the results of your decisions is the best teacher, so if you are not already keeping a "book" on each track, start one now. These records can be helpful when traveling to a new track that is similar to one you have experience racing on.

These three photos show how Scott adjusts the angle of his car according to track conditions. At the bottom, the car is very sideways when the track is wet and heavy. In the middle, the car is straighter as the track begins to dry out. On the top, Scott keeps the car very straight to best handle dry, slick conditions. The angle of the grooves must also be adjusted to best present the biting edges to the track surface.

Soft Tires

Racers rarely run soft tires in feature races, especially 100-lap events, but grooving soft tires correctly can be a big help in qualifying or short heats on a wet track. Softer tires are generally used on tracks that have a lot of moisture but not a lot of abrasion. You may also use a soft tire on a surface

The grooving iron on the left has the "square" cutting blade (nearly all square blades have a rounded lower end) installed normally. On the right, the same-style blade is installed upside-down to use the separate ends of the blade to cut sipes in the tire.

---

that does not generate a lot of heat in the tire and may have loose dirt or clay on the surface throughout the race. Cutting more grooves will help clean away loose dirt plus improve traction because of the increased number of edges available to dig into the track's surface. On tracks where you are not moving or throwing any dirt, but the surface is still relatively soft, a soft tire can often be run effectively with little or no grooving at all.

Keep in mind that grooving not only increases traction, but it also increases the rate at which a tire wears. A soft tire will begin losing its traction sooner as the amount of grooving increases. The softer material naturally tears easier and fatigues sooner, causing the tire to slow down. On a surface with a lot of traction or that contains rocks and other debris, excessive grooving can cause the tread to tear up much quicker. Determining the correct amount and style of grooving comes with experience, watching how the tire wears, and what the driver felt as the race goes on.

One of the things we do with soft tires is only groove about halfway across the blocks to prevent weakening the structure too much. We can often run the large stagger-block tread design with little or no additional grooving, depending on the condition of the track. Usually we will run that pattern on all four corners on a wet track.

Occasionally, when we run a ribbed tire on the right-rear and do the grooving ourselves, we will follow the groove lines the factory put on the tire and then wait to see if the tire is going to start tearing, and how many laps it takes for the tire to start working well. If the tire needs a couple of laps before it starts working, we may add some grooves or sipes to give the tire a little extra grab until it develops enough heat to work on its own.

Hard Tires

Even though some harder tires may withstand grooving better than the soft compounds, the track conditions that made you choose a hard tire may not require much in the way of grooving. In the South, we often do not groove the harder tires very much. In fact, there are some tracks that pack down to where they are similar to asphalt and we have run tires with no grooves at all. We leave the tires full-slick on a lot of the high-abrasion racetracks.

There are also hard-tire racetracks that have little to no abrasion or tendencies to build heat in the tire. On these tracks it may be necessary to groove a tire more than you would for any other condition. The extra grooving produces more sharp edges that grab the track surface, plus helps reduce the possibility of the tire glazing over and slowing down.

Sidewall Grooving

Grooving the sidewalls can be helpful if you plan to run a high line or on the cushion and need to be moving some dirt. Grooves on the sidewalls help clean away some of the loose dirt to get at moisture beneath it. Sometimes the outer row of blocks on the right-rear are also grooved to work with the sidewalls.

On some of the harder natural-rubber tires we sometimes sipe the sidewalls. This can really help when you are rolling the tire under when running lower tire pressures on a very slick track. The sipes can help prevent the sidewall area of the tire from glazing over and losing traction, just as on the tread surface. The sidewall is as important a part of the tire as anything else, and if you are going to be running on it due to low air pressure, or because you are running against the cushion, you need to make the best use of it.

Grooving Widths

Some tracks do not have a lot of abrasion to them, and edges on the tire can increase traction considerably. When we get on a surface like that, we will use wider grooves to present a cleaner, more prominent edge to the racing surface. Narrow grooves may not have enough distance between their edges to allow them to attack the track's surface properly.

On some tracks, you can groove the tires twice as much with a narrow groove or half as much with a wide groove, and accomplish the same traction, depending on the condition of the track surface and how abrasive it is, or if it contains rocks that will tear the tire up. Running a tire with a lot of grooves on a rocky or highly abrasive track could lead to the tire slowing down early or the tread tearing up. Fewer but wider grooves stand up to these harsher conditions better.

Grooving Angles

The angle at which grooves are cut determines how effectively the edges are exposed to the track when the car is in various degrees of slide. Dirt race cars seldom (if ever) run in a straight line, but the driver will try to keep the car much straighter when the track is slick than when there is a lot of bite to work with. When the track is slick, we keep the grooves nearly straight. But when we spend a lot of time with the car sideways we will put more angle into the grooves.

How much of an angle is dependent on the driver's style, and experience is the only way to determine the best angle for your situation. The idea is to keep the maximum amount of the tires edges facing the direction the tire is actually traveling. As a dirt car travels around the track, the body is pointed toward the infield, but the grooves must be angled to properly attack the track in the direction of travel. A consistent driver makes determining the right angle for the tire grooves that will work best for your combination much easier.

Circumference Grooves

We almost always cut circumference grooves in our front tires because they help make the tread blocks more flexible, which increases traction and helps the steering. Here, also, the width of the groove is determined by individual track conditions.

At some tracks we will also sipe across the

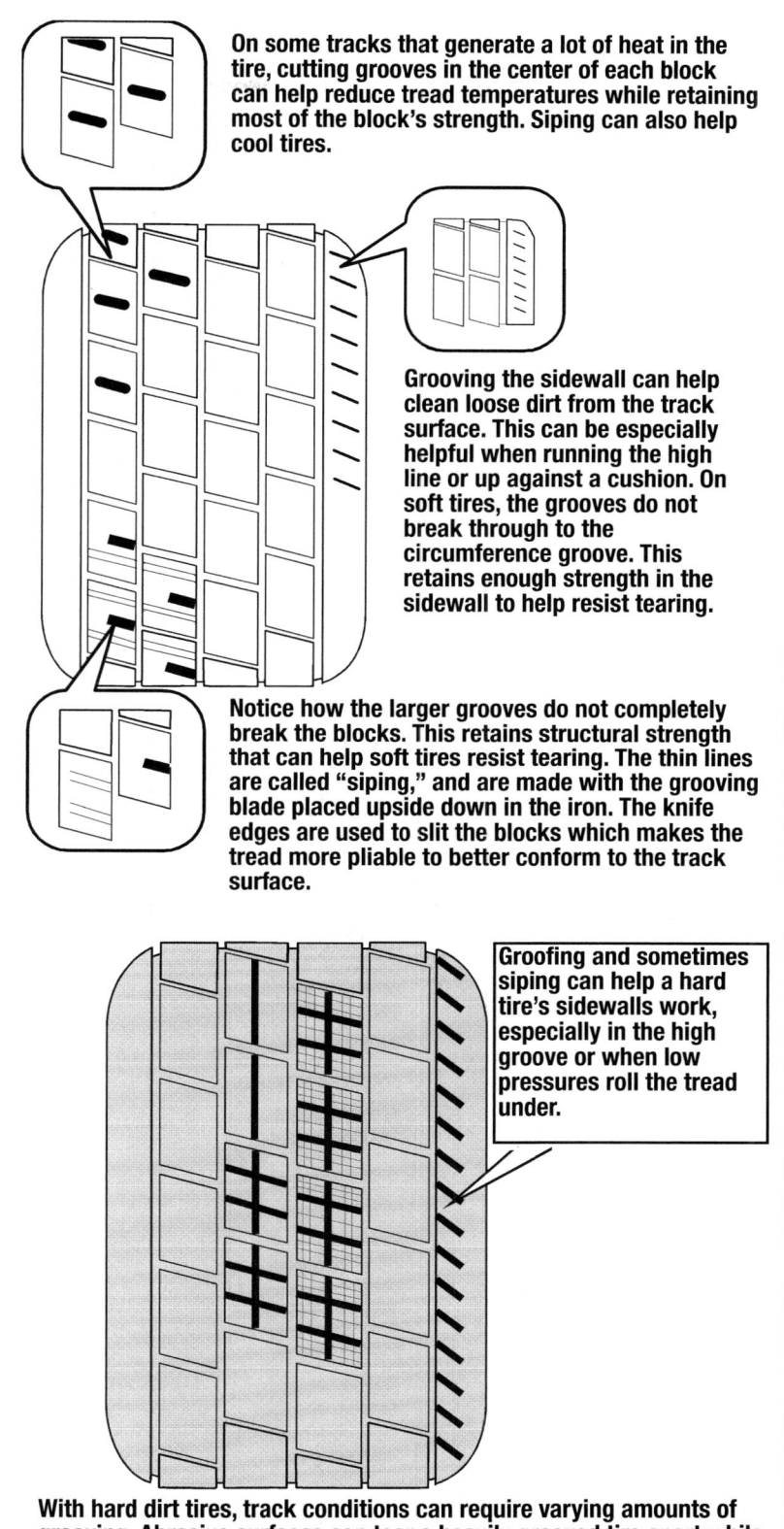

On some tracks that generate a lot of heat in the tire, cutting grooves in the center of each block can help reduce tread temperatures while retaining most of the block's strength. Siping can also help cool tires.

Grooving the sidewall can help clean loose dirt from the track surface. This can be especially helpful when running the high line or up against a cushion. On soft tires, the grooves do not break through to the circumference groove. This retains enough strength in the sidewall to help resist tearing.

Notice how the larger grooves do not completely break the blocks. This retains structural strength that can help soft tires resist tearing. The thin lines are called "siping," and are made with the grooving blade placed upside down in the iron. The knife edges are used to slit the blocks which makes the tread more pliable to better conform to the track surface.

Groofing and sometimes siping can help a hard tire's sidewalls work, especially in the high groove or when low pressures roll the tread under.

With hard dirt tires, track conditions can require varying amounts of grooving. Abrasive surfaces can tear a heavily grooved tire apart while less abrasive conditions need more edges to develop the most traction. Siping hard tires can be used alone or with other grooving techniques to make the tread more pliable and to dissipate heat.

circumference grooves so that we are making little blocks that are even more flexible. These smaller blocks can help maintain steering and braking response when you catch little rough areas on the track.

Rough Racetracks

I have found that as the roughness of a racetrack increases, so does the need for grooves. Unless the track surface is very smooth, some sort of grooving is likely to be necessary in order to be fast. When you groove a tire more, it makes the tread surface much more flexible. It can then follow or shape itself to the track surface much better to provide the maximum amount of bite possible.

Grooving and Heat

Some tracks naturally put a lot of heat into a tire and can actually cause the tire to melt or blister. On these kinds of tracks we have found that grooving helps cool the tire. The grooves (and sipes) help move air across the tread surface, keeping overall tire temperatures down. Grooving also produces more surface area, which helps transfer the heat out of the tire to the surrounding air. Racers have to balance grooving with the amount of abrasion or traction available, because tracks that generate a lot of heat may also be very abrasive or contain rocks that will tear the tire up if it is grooved too much.

To help a tire dissipate heat without excessively weakening the tread blocks, we will cut grooves in the center of each block instead of cutting all the way across it. We have even made a tool for drilling a circle in the middle of individual blocks to help dissipate heat without weakening the block's structure too much. This kind of grooving works well when you want as few grooves as possible for speed but need to increase cooling to prevent blistering.

Race Length

The number of laps to be run is an important consideration when deciding how to groove your tires. Softer tires are sometimes used in shorter races of 25 to 40 laps, but racers will have to be careful about over-grooving the tires. Removing too much rubber weakens the tread blocks and accelerates the rate of wear, which can cause the tire to slow down or fail prematurely. Accelerated wear caused by excessive grooving can also be a problem with harder compound tires often used in 100-lap races, depending on how abrasive the track is.

We try to decide how much wear we expect on a particular track, then adjust our grooving accordingly. Often we will groove only halfway into the depth of the tread so as the race progresses, and the track turns smooth and abrasive, the grooves become very shallow, or begin to disappear entirely as the tread wears away. Then, in the latter stages of the race, when we need the most rubber on the track, the grooves are just about gone.

KEEP RECORDS!

Learning the tracks on which you race and how they change over the course of an evening's racing will go a long way in helping to learn how to groove tires. Keep notes describing how your last grooving ideas worked and include the condition of the track surface. What were the other guys who drove past you doing? Read your tires after the race to see if another style or amount of grooving would have worked better. Did the tread tear up? Did you need more rubber and less grooving at the end of the race? Did the tires glaze over? Write it all down so you will have that information the next time you are in a similar situation.

Much of learning to prepare and drive on dirt surfaces comes from experience and applying that information to the ever-changing conditions of dirt track racing. You may have to spend some time writing down what didn't work but that puts you closer to what will work to make you faster.

Chapter 19

PRESSURES TO WIN

Set Tire Pressures For Speed

Text by Scott Bloomquist
Photos by Tom Hintz

Adjusting tire pressure is one of the arts in our sport that can be very difficult for racers to master. Some racers base tire pressure decisions on what they see being used by faster cars, without understanding how these changes affect handling or how to best use them to improve their car. Small changes in tire pressure can make the difference between a consistent, good-handling car and one that eats tires while running farther back in the pack than it should. The more we understand about tire pressure, the better we can use it to make our cars faster and easier to drive.

WEIGHT, WIDTH AND CONSTRUCTION

A tire is much like a balloon, the more air we put into it the harder it gets, and the smaller the contact patch becomes. This can be a big problem when racers from one division imitate the pressures used in other kinds of cars. Wider tires can support more weight at a given pressure than a narrower tire at the same pressure. A 3000-pound car on 8-inch tires simply cannot run pressures as low as a 2500-pound Late Model on 11- or 12-inch-wide tires.

With a Late Model dirt car, varying weight by 200 to 300 pounds doesn't seem to make much difference to air pressure settings. When we didn't have weight rules we often ran a little less pressure because the cars were so light. We also must remember that a lighter car generally goes faster and that extra speed works the tire harder by

increasing the centrifugal forces during cornering, which try to pull the tread under, especially if the pressures are set a little too low.

Sidewall construction is extremely important to how a tire responds to pressure changes. We use Hoosier dirt tires on our Late Model and have found the LSB (Large Stagger Block) tire has a stiffer sidewall that resists rolling under better than the economy ribbed tire that has a more flexible sidewall. When running the ribbed tire we often have to increase pressures by one to two pounds for the tire to work best. The pressure increase may seem small, but it has a sizable affect on how the tire works and how the car drives.

Tire height also has a significant impact on pressure requirements because of the leverage exerted on the tire as the weight of the car moves sideways in response to the centrifugal forces of cornering. Low-profile tires definitely roll under less, and on tracks that have a lot of traction, can help free up the car. They also may need lower pressures to work as well as a taller tire under the same conditions.

I've heard racers say they increased pressure to stretch a tire to get the right stagger, but what they are forgetting is that changing tire pressures by one pound can change the amount of bite by four to five pounds. If a tire is not the right size, you need a different tire to get the desired stagger, not more pressure. The same applies to reducing pressure beyond your normal limits to decrease its size.

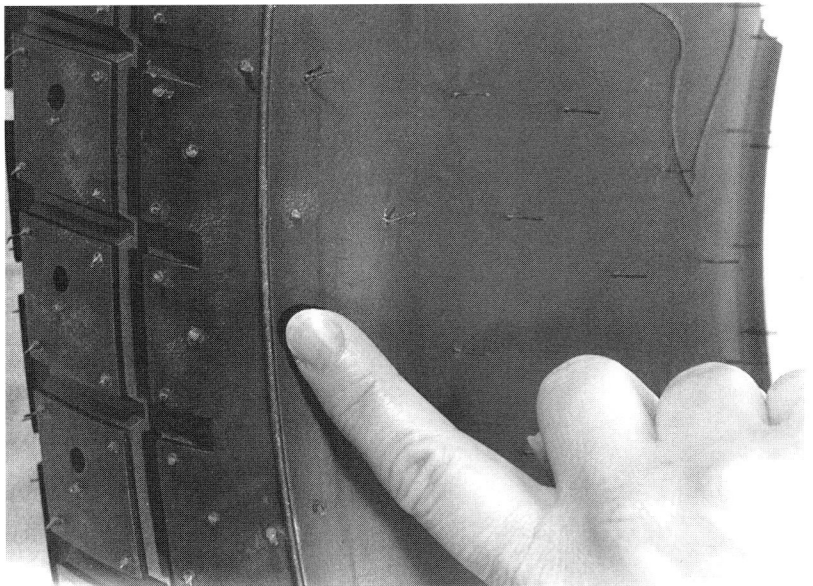

Most dirt tires have a line or other marking on the sidewall indicating where the thicker tread tapers down to the thin sidewall fabric. Underinflated tires can fold under enough to run the thin part of the sidewall on the track, greatly increasing the chances of a rock or other debris puncturing the tire.

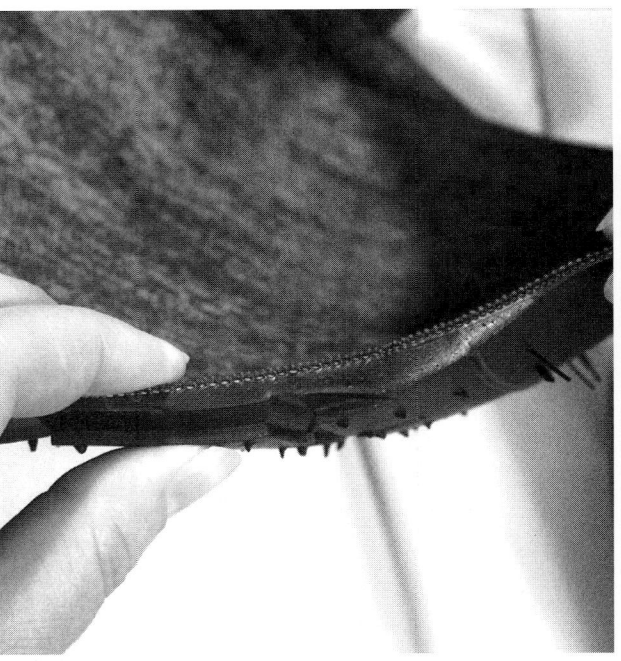

This cutaway section of tire shows just how drastic the difference in thickness is between the tread surface and the sidewall fabric. Look closely and you can see the line on the outside of the tire (near right hand) marking the end of the useable sidewall. If your tires are wearing above this line, you need more pressure.

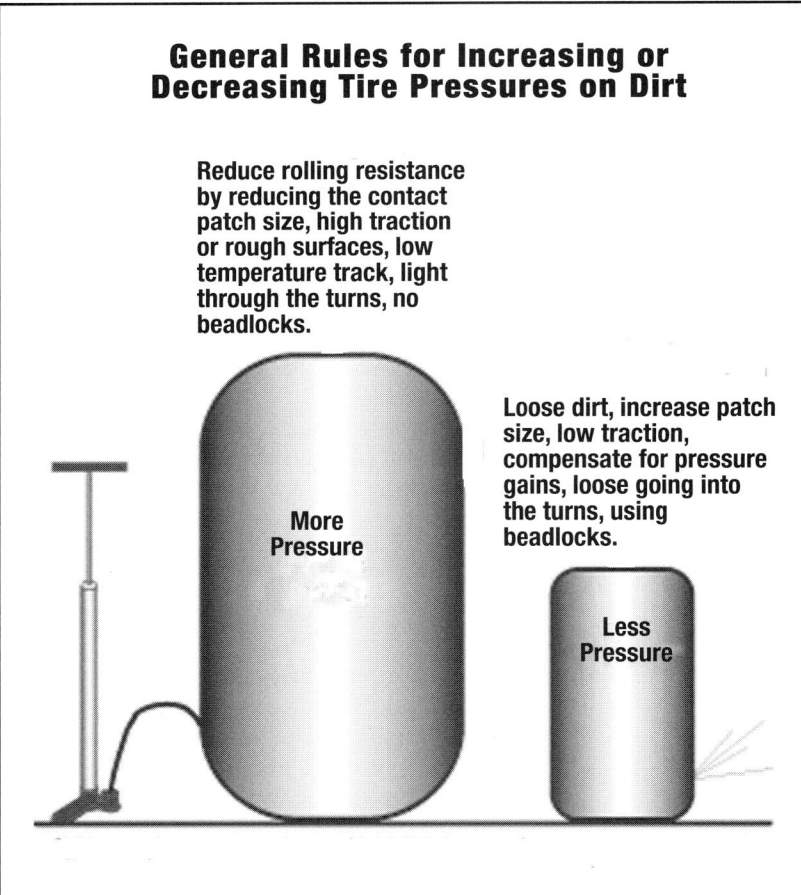

General Rules for Increasing or Decreasing Tire Pressures on Dirt

Reduce rolling resistance by reducing the contact patch size, high traction or rough surfaces, low temperature track, light through the turns, no beadlocks.

More Pressure

Loose dirt, increase patch size, low traction, compensate for pressure gains, loose going into the turns, using beadlocks.

Less Pressure

TIRE TEMPERATURES AND PRESSURE BUILDUP

Tire temperatures have long been a useful setup tool on asphalt but can be difficult to use on dirt tracks because there may not be enough heat generated in the tire to get accurate readings. But when tire temperatures get high enough we use them to help set caster, camber, and check the effectiveness of our tire pressures.

Temperatures are taken just as they are on asphalt, in three places across the tread of each tire. Generally when the center of the tread is getting hotter than the sides, the pressure is too high. If both sides are hotter than the center, the pressure is probably too low. If either the inner or outer portion of the tread (not both) is heating up, the problem is generally with the camber settings. Anytime one portion of the tire is working harder than the rest that section can fail much sooner than expected, and is likely to cause the rest of the tire to fail in the process.

Tires nearly always build pressure during a race, but some teams choose not to use relief valves. Pressure-relief valves work fine in many kinds of racing but we do not use them because of how some dirt tracks change over the course of an event. Sometimes tire temperatures will go up

TIRE WEAR INDICATORS

If the center of the tread wears or wears more than the sides, the tire is probably over-inflated. This can also occur when heat builds excessive pressure during a race.

When only the side of the tread closest to the wall shows increased wear, the pressures may be too low, allowing the tire to roll under. Another cause could be incorrect camber settings.

When both outer edges of the tread wear more than the center, the pressure is probably too low for the conditions, and should be increased for a more even contact patch.

pounds of air out to compensate before going out for the final 50 laps. During the second segment enough rubber was put down that the tires stopped spinning and tire temperatures dropped, lowering my right-rear to eight and a half pounds.

Atomic Speedway is a very fast track that often needs a little higher pressure to be fast, and lowering the pressure made our car very unstable and inconsistent, and it never felt right throughout the rest of the race. If we had left the pressures alone, the cooling effect of the track late in the race would have brought us right back to where we originally wanted to be.

On tracks where we expect pressures to go up substantially, we may run Nitrogen rather than normal air in the tires. Nitrogen does not expand as much as normal air when subjected to heat and helps prevent running on the center of an over-inflated tire later in the race. Although Nitrogen helps control unwanted expansion, it will not correct improper pressure settings.

The longer the race, the more we can expect heat and pressure to build up, but keep in mind that the majority of the increase occurs during the first five or 10 laps of hard racing. One way we limit the amount of pressure buildup on high-heat tracks is to lower the initial pressure an extra pound or so, especially in the right-rear. Then during warmups we try to heat up the tires a little to gain that pound back in order to make the car handle well at the start of the race. The tire still builds pressure, but lowering the starting point reduces the maximum pressure developed.

substantially early in the race, but later the track conditions may cause them to cool off again. If the relief valves dissipated the built-up pressure we could find ourselves with under-inflated tires when temperatures go down.

Knowing what the track is likely to do later in the race is an important piece of information to consider when setting tire pressures. For example, we used to run some 100-lap races in 50-lap segments and got ourselves in trouble adjusting air pressure during the break at Atomic Speedway (in Tennessee).

Early in the race the track was slick and let the tires spin, which built a lot of heat in them. Our right-rear built up to 16 pounds, so we let four

Check and record the amount of pressure your tires gained immediately after each race, and include a description of the track conditions. As you accumulate this data and experience using it, accurately predicting how much pressure gain to expect in similar situations will become much easier.

PRESSURE POINTS

Craig Cowen of Hoosier Tire has seen Scott Bloomquist win a lot of races and thinks his use of tire pressures is one of the ways he does it. Cowen reports, "Scott often runs different air pressures than a lot of the other guys. I've seen many races where he has run considerably more air pressure than the next guy. Part of this is because of his many different styles of driving, but Scott also knows enough about tires to use pressures to fine-tune his car for the conditions."

Red Kosinski is one of Goodyear's oldest racing tire distributors and has been around dirt racing for more than 30 years. Kosinski states, "We see a guy lose a tire and he can't figure out why. It's often because they are running the tire pressures too low, thinking that will give them more bite, but it really doesn't. You have to run the right pressure to keep the tire up and on the rim. If it falls off the rim you lose. That is the biggest mistake racers make."

While tire companies cannot make broad recommendations as to what tire pressure will work in all situations, Kosinski has some simple advice. He instructs, "You have to run sufficient air pressure in the tire to prevent it [from] tucking under. If you start running on the sidewalls you're not accomplishing anything. There is no compound on the sidewalls, it's only on the tread surface. The sidewalls are made of nylon that was not designed for traction."

Kosinski feels the Street Stock level racer must run higher pressures because of the weight of their cars and the lack of sophisticated suspensions. He relates, "These are not high-dollar race cars. They don't have the trick racing shocks and springs to keep all of the tires on the track. When they dive into a corner the whole car unloads onto the right-front tire momentarily. That's when the tire folds under, hits the suspension, and blows out or peels off the rim. I tell these Street Stock drivers to run 30 pounds of air in the right-front tire and then work it down gradually from there, in two- or three-pound increments until they find the right pressure. Everybody thinks beadlocks are the answer, they do help, but you still have to keep the tread on the racetrack."

Track Surface

The composition of a dirt track, as well as the amount of available traction and how smooth or rough the surface is, all play a role in setting tire pressures. High levels of traction allow you to get into a corner faster, but that increases the lateral loads on the sidewalls, making higher pressures necessary. On tracks with a lot of loose dirt we may lower tire pressures to increase traction, keeping in mind they must remain high enough so that the tire does not distort excessively, which can lead to it failing or rolling off the wheel.

When a track is very fast with a substantial amount of traction available, a little more air pressure is likely to be faster. Higher pressure reduces the size of the contact patch (less friction), so the tire rolls more freely and generates more speed, plus it can help free up the car. When the track is slow and doesn't have a lot of traction, you may find yourself lowering tire pressures to increase the size of the contact patch and get a better grip on the track.

All drivers have individual tendencies and styles that influence what tire pressures will work best for them. Drivers who try to keep the car very straight may be able to use a little lower tire pressure.

Drivers who turn the car hard or throw the right-rear up against the cushion are going to need higher pressures to handle the forces of running against (and into) the berm, or they could be parked for the night.

The difference between the correct tire pressure for the conditions and running tires too low or too hard is often very small. If the tire pressures are a little high you may give up some traction, but run them too low and an overheated or blown tire could take you out of the race altogether. I don't think there is any speed in excessively low tire pressures but they can definitely hurt performance. If you find yourself running tire pressures lower than normal to increase traction, there may be a problem elsewhere in the chassis that should be corrected instead.

Look over the racing surface for rocks, bumps and holes, then keep an eye on how the track changes throughout the event. If a track is known to develop big ruts or a sizable berm you may have to increase tire pressures accordingly. There may be corners that you want to run into hard, but if the tire pressure is too low the sidewall can fold under, letting the rim impact the sidewall and the ground and possibly cutting the tire. Later, if the track slicks-off, lower pressures may help develop more traction if you can avoid hitting big ruts or the berm. In some cases pressures will have to be raised to survive a corner that is particularly rough.

WHEN IS THE PRESSURE TOO LOW?

Even though lower tire pressures increase side bite, when you get too low they will start loosening up the car through the corner, and being too low can possibly beat a tire to death. Insufficient air pressure can cause the tread and/or sidewall to develop a wave that over-flexes the tire and causes it to build a substantial amount of heat. This exaggerated flexing can break down the tires construction, causing leaks through the shoulder between the tread and sidewall, or a complete failure of the tire. We've all seen the guy who has a tire going down but continues racing. When the pressure gets too low the flexing builds extreme temperatures and the tire fails. The combination of excessive temperatures and damage from over-flexing can get so severe that the tire may literally blow apart.

One of the keys to setting tire pressures is to keep an eye on the sidewall. There is a certain amount of the sidewall that you need to be working every time you race. On most tires there is a line where the workable sidewall ends and it turns into a much thinner layer of rubber and fabric. Running on the fabric portion of the sidewall drastically increases the chances of a rock or other debris cutting down the tire.

With the tire dismounted, you can feel where the thicker rubber tapers off and the thinner fabric portion of the sidewall begins. There is no portion of the thin sidewall that can be run on safely. Some tires have as much as 1 1/2 to 2 inches of the heavier sidewall from the last groove that is safe to race on. With this in mind, we have to set tire pressures high enough to prevent rolling the tire under beyond that point. Our team inspects the tires after each race to see how much pressure they gained and how much of the sidewall we were running on. Then the pressures are adjusted to keep the tire working on the tread.

An easy way to see how much of the sidewall is being used is to paint a wide stripe on the side of

Even the tread surface itself is not as thick as a street tire. When any tire is run with insufficient air pressure the tread surface can form a wave that over-flexes the layers. Over-flexing causes excessive heat and can separate the layers, causing everything from air leaks to a complete failure of an otherwise perfect tire.

the tire facing the outside wall, extending from the rim down onto the tread. Use shoe polish or latex paint to avoid attacking the tire itself. A few consistent laps will wear the stripe away where the tire is contacting the track surface.

Paying close attention to how a tire has worn after the race is another way to determine how well you are working the contact patch. Just as with temperatures, the concentration of wear or tearing will show how effective your pressure settings are for the conditions. If the outer edges of the tread are tearing or wearing more than the center you need to increase the pressure. If the center is wearing faster than the outer edges the pressure has to be reduced a little. If only the outer edge (facing the outer wall) is wearing, the tire may be rolling under because the pressure is too low. The changes needed to correct the wear are usually small and should be made gradually until you get a feel for how much of a difference one pound of air makes in your situation.

Air Pressures and Spring Rates

On dirt we seldom find ourselves having to work with air pressure the way we work with springs and shocks for handling. Although air pressure changes affect the spring rate of a tire, they are more often

Veteran dirt Late Model driver Ronnie Johnson sets his tire pressures just before going out for 100 laps against some of the best Hav-A-Tampa drivers in the country. Johnson finished fifth and his team owner, Scott Bloomquist, won the race, leading virtually all 100 laps.

used as a tool to get the optimum traction for the condition of the racetrack. If large tire-pressure changes appear necessary to get the car working, consider changing the spring rates or shocks instead.

The key to winning races is consistency, and proper tire pressures help prevent radical changes in how a car drives throughout a race. Manipulating tire pressures can help get your car hooked up a little better than your competitors. Tires will last longer and work better with correct pressure settings, which will make you faster in the later laps of the race.

DIRT CHASSIS
DRIVING TIPS

Chapter 20
HOW TO READ A DIRT TRACK

Text by Scott Bloomquist
Photography by Tom Hintz

Scott Bloomquist is one of the most successful dirt racers in history. You're going to be faster on dirt after reading and following his advice.

The one point all dirt racers have in common, regardless of their division or level of experience, is the consistency of the surface they race on. Each driver must contend with the same dry spots, bumps, ruts, and holes encountered during a race. Dirt tracks change from day to day, from qualifying to the heats, and often several times during the feature. Drivers who identify and deal with problem areas and find the fastest line around them greatly improve their chances of winning the race.

IDENTIFY THE DIRT

Probably the most important part of reading a dirt track is determining the composition of the racing surface. At one time I didn't know how important this was, but it is probably the most important item to know. Drivers have to learn to recognize the different kinds of dirt, the amount of sand and rock it contains, and how that affects the track during the course of a race. It seems like each state has its own kind of dirt. Occasionally there are similarities, but most of the time the dirt is very different. Some kinds of dirt put a lot of heat in a tire and will affect your choice of compounds. With experience, you can learn to recognize the difference in practice by taking tire temperatures.

TRACK PREPARATION

Usually, by the time we arrive, most of the track preparation has already been done. I ask around the pits to see if the track has been dug up, and if it has I ask if it was dug up very deep. You can almost count on a track that has

been dug up several inches deep, and then regraded, to take a softer tire. A track that has been raced on week after week without being torn up between races will continue to get harder and harder regardless of how much water is dumped on it.

Racers really have to go out to see how soft the track is for themselves. No matter how it has been prepared, you have to get your own feel for how soft the surface is, and apply that to what you look for and how you plan the line you're going to drive that night.

Take a screwdriver or something similar and see how far you can penetrate the track. You can feel the changes in consistency of the dirt from where it was dug up to where it is still hard-packed. If you can push the screwdriver way down into the track, you can count on it bleeding moisture back later in the night. If you can only penetrate a little bit, then you know the track hasn't been dug up, and you will have to choose tires accordingly. Just don't use something real sharp, such as a knife, that will penetrate regardless when checking the track.

Drivers also have to know if a track has been treated with a sheepsfoot roller. Nine times out of 10, sheepsfooting creates a harder, slick surface, and reduces the amount of surface for your tire to grip on. Generally a track that has been sheepsfooted will be slower, and get harder and drier quicker.

Drivers also have to pay attention to where the water is put down. On some Southern tracks, if one part of the track gets less water, it has a tendency to get very dry and abrasive and will get fast in that spot. In the North, the dry spots will

Notice how two distinct lanes are being developed on this dirt track.

usually be the slow areas, and you won't want to run on that part of the racetrack. Much of this has to do with the kind of dirt the track is made of. If you know what kind of dirt is at a certain track, you will have a good idea of where on the track you will be running, and what kind of tire you will need.

SELECTING TIRES

Learning to effectively read a dirt track is crucial to being successful in 100-lap races. You absolutely have to pick the right tire to run that long and be competitive at the end. Local racers often wonder how traveling teams can come into their home track and beat them in a 100-lap race. So many of the local guys know what it takes to run their weekly racetrack for 25 to 40 laps, but then you have a lot of cars that come in for a 100-lap event and the outsiders beat them in the end.

The problem is that in a 25- to 40-lap race, the track goes through two or three changes, but in a 100-lap race, it may go through another three or four stages before it reaches how it's going to be at the end of the race. The local guys will often qualify up front, but they won't make the right tire selection for the long duration of the race. We have a lot of experience picking tires for the long races and that helps us finish ahead of the local drivers. If the races were shorter, the results would probably be a lot different.

One of the ways I check a track for adhesion is to walk out and scuff my driving shoe on the surface. After you do that for a while you start learning what to feel for and what that feeling is telling you.

If I go out and slide my shoe across the track, or even kick at it, and the shoe sticks, or makes a squealing sound, I know this track is going to take some hard tires. When my shoe feels almost like it has baby powder on it when it slides across the track, I can feel there is less adhesion, and we will need softer tires. Getting the feel for this may take a little practice, but if you remember what the track felt like to your shoe and how the tires worked later that night, you start getting the idea and will be better able to pick the right tire for the race.

Another very important part of reading a racetrack is a stopwatch. A person can probably learn more from a stopwatch than they realize. If lap times are picking up or slowing down, you can tell a lot about where the racetrack is headed. If you keep track of how lap times are changing and relate that to how the track felt during your races, you will see that a stopwatch is an important tool for picking tires and predicting track conditions.

HOT LAPS

Hot-lap sessions can tell a driver a lot of information. You get to check your car to make sure there are no obvious problems with engine response and that the chassis is basically working. The most

Ronnie Johnson came out onto the track to check the surface before making final adjustments and tire selections for the night. Ronnie and Scott Bloomquist were the only drivers we saw doing this. Ronnie won with Scott right on his bumper at the end of 100 laps.

important information is on the track itself.

At almost every track there is a main groove around it that 90 percent of the cars are running. I always look for a way to make my car handle in a line that the other drivers are not running, because there will be moisture left there after the more popular line dries up.

Most of the better racers will hunt for their own line around a racetrack, which will give them the most traction. I've been to a lot of racetracks where everyone will run in the middle of the track. It gets black and worn out, but there is moisture right at the bottom and right at the top.

Sometimes you can enter a corner extremely high, use that little bit of bite that is way outside, and turn the car extra hard to come across the black (middle) line to the bottom during the exit to hit the moisture there. You use the outside moisture to get across the slick area instead of staying in it all the way through the corner. I've been on the pole every time we've gone to Cleveland (Ohio) and have been able to do that by driving this kind of line.

Hot laps are where I really decide on my qualifying line. I try every line possible just to be sure my car is capable of running that many different lines and of being fast in all of them. If you get a bad number (qualifying order) and have to qualify late when the track is worn out, you want to know your capabilities—whether you can run an extremely different line while hunting for bite that could still give you fast time.

WATCH THE TRACK CHANGE

I think about 80 percent of reading a dirt track is visual. You can see it change probably easier than you can feel it. I'm constantly looking for something on the racetrack—the dust or the moisture—that shows me another line to run.

There is a phenomenon on a lot of tracks when they get dry, black, and slick where all of a sudden you see a red stripe come back. This is happening because the cars are wearing rubber the tires laid down earlier back off the surface. Often the track will pick up and get fast again—really fast through that line. You have to keep an eye out for that, because when it starts happening, the track starts picking up and getting faster and faster. If you get out of that area you can leave the park. This stripe can make the track 1- to 1 1/2-seconds faster, and if you get up out of that bite, there's not enough traction to hold the car on the racetrack.

Drivers should also remember how much moisture there was high in the corners when you first got there. So many times the cars run lower on the track and throw dirt and dust into the upper groove, making it hard to get up there and run. But if you know there is a lot of moisture up there, later in the race you can go up, run that line a couple of laps, and blow that loose dirt off. You may have to sacrifice a position or two to get that line run in, but you might develop a really good groove for yourself.

We did exactly that at the Pennsylvania Motor Speedway. We came in, changed tires, and everybody was running the bottom two and three lines, but the track is five lines wide. We went out on the high line, blew the dust off, and found some bite up there. Using that line, we came from 30th position to the lead in five laps.

The potential for finding good bite is often there, you just have to be willing to try it. Sometimes guys are too reluctant to lose their fifth position to try anything. If you have the right attitude, you look for a faster line. If you're not in the lead, you need to be doing something different. You need to either run a line in, or hunt one up that will make the car stick and get you to the front.

PREDICTING RUTS, BUMPS AND HOLES

Some tracks commonly develop rough spots that you want to watch for, but other than that, it's really difficult to predict where a track is going to rut up. Generally, a rough section of a track is likely to get rougher as the night goes on. Once the cars start bouncing through bumps, they will make more or deeper ruts and bumps. If you see an area

In hot laps, Scott Bloomquist ran every conceivable line to see how his car performed in each of them. During the feature, we noticed him hunting for those lines again until he settled on one that worked the best that night. He nearly won the race, finishing a very close second to Ronnie Johnson, who also moved around searching for a good line during the race.

that looks like it has been rough, go out and stick it with a screwdriver. If the track is soft, you can count on it getting rough again. If the surface is very hard, it might hold its shape through the night. Either way, if you can get your car working in a line that misses those areas you could have an advantage later in the race.

CUPPING AND CROWNS

When you first go out on a track in a race car, you may see where the track is dished out, or cupped. That tells you where most of the cars have been running and that the track may not have been dug up recently. That may be the only fast line around some tracks and you will have to look for the fastest part of that line.

Something else you may see on a track that has not been dug up often is "crowning." That is where the dirt is slowly moved up the track and forms a rounded hill or crown near the top. The low side of this crown can be fast, but moving up another foot could put you over its center onto the downside of the hill and into the fence very quickly.

READ A TRACK AND KEEP RECORDS

Most of us have to learn to read a track through experience. Keeping records of what you see and feel in the car will help teach you to predict where the best lines can be expected later in the race. Keeping a book with information on a track you

run a lot or plan to come back to is a big help.

When you get to a track, go out and stick the surface in a lot of places, from the bottom to the top. Remember where it was soft and hard and where the most moisture was. Then compare that to how the track turned out at the end of the night, where most of the cars ran, and where your car was fastest. Then go out and stick the track again to see how much and where it changed. Keep records of what tires you ran, what was successful for your car and what wasn't. Then connect that with what you did with tires that evening.

Keeping track of the weather along with the other information is extremely important. Weather has a big effect on how a track changes over the course of an evening. A cloudy day with some light showers can put a different face on a track. Wind is especially tough on the moisture. Add a hot sun to the wind, and it's hard to get enough water on a track to keep bite in it for long.

Learning to read a dirt track takes practice and experience, just like learning to drive a race car. Keeping good records helps a lot and can shorten the learning time considerably. Some of the drivers passing you late in a race may not always be a better driver or have a better car. But they probably know more about what the track is doing and where to find the bite to get by you. This kind of speed doesn't cost you anything but your time and effort.

Chapter 21
QUALIFYING

The Keys to Starting Up Front

by Ronnie Johnson with Nick Masters
Photos by Nick Masters and Tony Hammett

Ronnie Johnson on a qualifying run at Atlantic Speedway for a STARS-sanctioned event dubbed the Dogwood 100—an event he would go on to win.

These days, Late Model racing has become ultra-competitive, with bigger and bigger fields trying to make the 24-car starting lineup. Obviously, qualifying, or a time trial as some people call it, has become more important than ever in giving yourself the best chance of making the big show. There are two main things to look at when you get to the track: 1) the driver and how he drives the car, and 2) the tires. There are all sorts of little technical things you can look at—smaller fan, different air filters, gas or alcohol, and so on. All kinds of little things that might save you one hundredth of a second and may even make a difference of 3 or 4 hp on the dyno. You can go to great lengths to get all of these little mechanical advantages, but if the driver doesn't do his job, or if you have the wrong tires, you are in trouble!

Also, time has become very important. Between hot laps and qualifying, you don't have the time or the manpower to change things and then turn around and change them back.

First, I try to watch the other guys' hot lap. I don't always get to do that, and I do it even less racing my own car. If we have problems, then I have to help take care of them. But a driver should watch the others with a stopwatch in hand, or have someone in the crew or a friend watch hot laps and get times on the three or four best drivers in each session. That way, if you are at a Hav-A-Tampa race where you only get one hot lap session, and they have four or five groups, then at least you have times on the top 10–15 drivers.

I try to get to a place where I can see the whole track, or at least a part of the track that has been a problem area for me. I'll either watch guys who usually run well on the track we're at or someone who runs well in the conditions we have, whether the track's fast, slow, wet, or abrasive.

TRACK CONDITIONS

If you have 60 to 100 cars, the track can change a lot in the period of time it takes to qualify the cars. If you go to Eldora for the Dream or the World 100 where they have maybe 200 entries, the track may change. But sometimes they qualify all those cars and the track doesn't change that much. So, you are still going to do the same thing whether you were the first or the last. At Eldora, it's far more important to make a good lap for those two races, because they only give you one lap at a time.

You need to be watching the racetrack—how it's changing. You need to keep in mind the number you have drawn to qualify and what the procedure is for qualifying. If there is no drawing system and the track is going to be faster early on, then you need to have everything done and be ready to get out on the track as quickly as possible. Be in line early and ready to go for a wetter-type racetrack.

For an abrasive surface or a sandy track, you need to be the last man out, not the next-to-last, but the last man out. Every car that goes out is going to leave more rubber on the surface of the track, and rubber-to-rubber contact is what gives you the most grip. That's why the rules were changed from going fast and taking only one lap if you missed your turn in the line. That waiting would mean you had an immediate advantage, even if you only got one lap.

Be ready for the conditions when it's your turn to go out. Look at the lines the other drivers are running. A stopwatch will tell you a lot, but sometimes you can just see it. A driver will be visibly quicker than anyone else through one part of the track, and he may be running a little bit of a different line than the others. There may be moisture down low or a cushion up high. You have to watch the track and not just respond to what the car does. You have to be able to look ahead for moisture on the racetrack or in places where there is more traction. This can give you a huge time advantage. If you are running softer tires, then you must find those spots and take advantage of them. This thinking even applies on the harder, more abrasive tracks. You must keep concentrating on the track surface to be able to see the places where your tires are going to serve you best.

One of the key things to remember is that you are racing the clock, not the other drivers. To prepare myself mentally, I sit in the car while waiting in line and try to pay attention to what the other drivers are doing. When I hear a time, I think to myself, "I can do better than that; I can go quicker than he can." You know that you and the crew have done everything possible, so you have to build on that to increase your confidence level.

TIRES

Your tire choice is very important, whether or not there is a tire rule. If there is no tire rule, then you need to know what everybody else is running. That's becoming harder and harder to do unless you have a good relationship with the other drivers. You can't always trust what you are told by other competitors, not because they would deliberately tell you something wrong, but because they might not always tell you everything. Drivers will usually tell you just enough so they can come and get information from you without worrying too much.

You can often tell what tires are on someone's car just by looking from a distance. It's not a good idea to just walk up and start crawling all over a man's car, unless you have that sort of relationship with him. If it's a new tire, you can tell by the tread pattern. If it's a used tire, you can tell by the way it has grained, or how it's grooved or siped. If you are used to handling tires, a lot of times you can see those things without having to walk up and look at

a compound number. Also, I have seen drivers change the numbers on tires just to try and fool the competition.

In qualifying, you have the whole track to yourself. At some tracks, the line that is being run is pretty well defined; and, unfortunately, most of those sandy abrasive tracks will only have a car-width groove in them. Tracks like that are the easiest to qualify on because the groove is so narrow that you have to run on it to get anywhere. Tire choice is usually obvious because you are limited as to what tires will work.

At some races you'll see a lot of different tires. You might have a guy qualify on #03s and get the pole. A guy might line up beside him having qualified on #55s, one of the harder compounds Hoosier makes. But the #55 is a great all-around tire that will work in a lot of different conditions. The guy with the softer #03 will have gone out and run around the bottom, taking advantage of the moisture with the left-side tires and keeping the car straight and smooth. But the #55 allows you to take advantage of the slightly rougher spots on the track and drive harder into the turns, being more aggressive and maybe using the cushion at the top of the track to help you turn.

Tires get hotter and therefore softer as you use them. It's good to have a chart for tires and tire use. I've seen Pup (Jimmy Thomas from Hoosier Tire South) try to provide that for people. There are too many variables to be super accurate. Even with experience it's still a guess, albeit an educated one. A basic chart that shows the beginner where to start would be a good thing.

The right-rear takes the most abuse and is the most important tire to get right. Depending on how you choose to set the car up, you can go out and try to make your first lap the best. Or you can try to get the feel of the track, get the tires to work under you, and then try to make your best lap. But if the track has a little moisture in it, or has started to slow down, or is a little greasy, and you chose to go out on a softer tire and try to hunt the moisture, then you have to make that first lap count.

Usually, we are racing in the summertime, and it's hot. So, even when you're qualifying, the soft tires will go off by the second lap. Sometimes in hot weather you can run a harder tire, which will do pretty good on the first lap. By the second lap it will have some heat in it and will actually give you a faster lap on the same line because it has more grip. At tracks that are more abrasive—those in coastal areas—a softer tire can be a disadvantage. At these tracks, you need a harder pavement-type rubber. It won't work well on the first lap, even if you have the luxury of being able to preheat it. You might have a decent first lap and pick up as much as one-half of a second on the next one. The first lap might not get you in the show, but the second might give you a fast time.

SETUP

I usually arrive at the track with the car pretty much set up the way it should be for qualifying. We really don't have that much time to change things once the program starts. You should use a standard setup that you know usually works well, unless you're going to a track with some different characteristic. For instance, some tracks are extremely fast, and you know that you are going to have to arrive with the car loosened up a bit. Usually you know that before you get there. But, sometimes you can go to a track like that and be too loose when you get there. You look around and see a guy who's never been there before and doesn't know any better, and he's flying. Ordinarily, as far as setup goes, I try to make my car a little looser or a little freer for qualifying than I would for the race. You might not have as much fuel in the car, maybe a little less tailweight or run a little more stagger. You are going to have the whole racetrack and may need to be able to move around on it.

The United Dirt Track Racing Association (UDTRA) body rules limit what you can do with the car's appearance for qualifying or racing. On a fast track, you can lay the spoiler down a bit for an increase in speed. More often than not, you will be a lot slower during the race than you were during qualifying. On the sandy and abrasive tracks the opposite can be true.

BREAK EXPERIENCE

I used to depend heavily on past experience, especially when going back to a track where we were successful, but not any more. We found that even if you are going back to a track for the umpteenth time, what worked before might not work this time.

With larger fields of cars, added prestige, and higher purses, it seems that track preparers and promoters are trying harder to make track surfaces better and more raceable. The guys who race weekly at a racetrack and know what's happening are now having a harder time when a big show comes to town. You'll have 15 guys come in who have never even seen the place before, and all of a sudden you qualify behind them, and you might be track champion. It isn't that they have better equipment, more money, or are better drivers. It's just that the track conditions are different, and you are still doing what you did last week. The conditions are rarely the same as they were the last time.

We aren't yet where we should be, but a definite change is taking place. You have to remember that we drivers are spending money to earn money. The only reward for the fans spending their money is to see a good race.

You don't need to throw away your notes and forget what you've learned. But you also don't need to sit around and wait for those same conditions to come around again, because they might not. You need to keep an open mind and be aware of what's going on all the time.

Chapter 22

ALL-OUT DIRT TRACKIN'

Billy Moyer Explains Dirt Driving Basics

by Billy Moyer and Tom Hintz
Photos and Illustrations by Tom Hintz

Billy Moyer knows how important it is for the driver to know what the car is doing and how to drive around problems. He works closely with his crew to get the car handling as best it can to reduce the alterations he has to make to what may be the fastest line.

Learning to drive a Stock car on dirt can be frustrating for new drivers who do not have a basic understanding of the techniques necessary for going fast on an ever-changing racing surface. Racing on dirt requires a flexibility in controlling the attitude of your car, while constantly changing your line through the turns. Pavement racing seldom requires a driver to make use of the often small areas of traction that dirt track racers must master.

Keeping the car as straight as possible will generate the most speed, but a wide range of variables, including the condition of the racing surface, missing your chassis setup, and traffic could mean sliding the car at varying angles is the best way to go fast that day. Drivers that are consistently fast on dirt know how to recognize where the best traction is, then how to change their driving to make the best use of it.

THROTTLE, BRAKES & THE SURFACE

In most forms of dirt racing accurate and predictable throttle control is at least as important in making a car go fast as nearly any other factor. IMCA Modified–style cars and dirt Late Models are surprisingly similar in this regard. Both usually have more power than their tires can successfully transfer to the track if the driver is too aggressive with the throttle. Learning how much throttle the track surface and your car are capable of handling that day will win races.

Whether driving in traffic or in the open, a driver has to be able to put the car on the best available line. When you are alone you can put the car where the best traction is. When you are in a crowd the ability to accurately change your line can make the difference between running behind everyone all night or getting by safely.

When driving against the best drivers in dirt racing, two cars can sometimes fit into one line. Having confidence in yourself and the other driver's ability makes racing this close possible.

It is very important to have a smooth working throttle (see sidebar page 117) with enough throw to be comfortable and controllable. While I like a relatively small amount of movement at the gas pedal, that can be a problem for some drivers, especially newer ones. Many feel a longer throttle throw at the gas pedal is easier to control and makes maintaining the right amount of power easier, especially on a rough track that bounces the driver around.

STEERING RATIOS

With the wide range of rack and pinion units available today, drivers can tailor their steering to the way they like it. Rack and pinion manufacturers have ratios from 3.4:1 to 1.75:1 (inches of wheel movement per revolution of the steering wheel) available, and one of them should fit your driving style. If you use a factory-style steering box it is equally as important to get a ratio that suits your style and experience. Factory-style boxes are generally slower than rack and pinion units but can be tailored somewhat by finding one with a ratio (or installing gear sets) that is comfortable for you.

Experienced drivers, those who stay calm in the car, and sometimes drivers with go-kart experience may adapt quicker to 3.0:1 or 2.5:1 ratios. Newer drivers and those who get pretty excited in the car usually need a slower steering ratio in the 1.75:1 to 2.0:1 range to help them keep the car under control. If you find yourself darting back and forth or feel like you often over-correct, you may need a slower, more forgiving steering ratio.

STEERING & SPEED

Turning the steering wheel creates friction as the front tires try to change the direction of the car's weight. That friction causes the car to slow down, or at least lose momentum. Even if you have enough horsepower to compensate for the additional friction you are squandering power that could be making you faster.

Ideally, race car drivers want to turn the steering wheel as little as possible. If your chassis is right for the conditions and you use the right amount of brake and throttle to hit your line, very little steering input is needed. I seldom turn the steering wheel more than 10 degrees to the left when the car is set up right for the conditions. In those cases I usually don't have to turn the wheel back past center (to the right) to get off the corner, which keeps the chassis free to go as fast as possible.

There are tracks and situations where the back of the car is going to be hung out to some degree and you will have to counter-steer to the right to maintain control. The key is to keep the slide to the minimum. Years ago everybody slid the cars almost completely sideways into the corners, but with the suspension technology we have today that just isn't the fast way around the track anymore. Even the Sprint cars who used to go into the corners almost backwards are now driving much straighter and trailing the brakes to pull the car down to set it. Sliding completely sideways and throwing dirt all over is great for PR pictures, but it will cost you positions or a win in the race.

THREE-WHEEL BRAKES

On wet, heavy tracks with a lot of traction many drivers like to use what we call three-wheel brakes. A valve is plumbed into the brake lines that allows the driver to reduce or eliminate braking at the right front wheel. When the brakes are stabbed entering the corner the braking is concentrated on the left side and the car turns into the corner very hard, swinging the rear end out at the same time.

There are a number of different style brake valves available. Some shut the braking off completely, while others allow the driver to adjust the amount of braking power delivered to the wheel being controlled by the valve. For most newer drivers, using an adjustable valve and gradually increasing the amount of braking removed from the right front wheel will help prevent a crash or slow lap times.

New drivers should be careful using three-wheel brakes, especially shutting the right front off completely, as the response to hitting the brake pedal can be very sudden. Here again, seat time will help you decide when to use three-wheel brakes or how much to reduce braking at the right front.

On many dirt tracks the surface has traction available all over it early in the evening, giving drivers the widest choice of lines when passing. There will still be one "fastest" line, depending on how well your car is setup.

DRIVING & CONDITIONS

Most dirt tracks dry out considerably during a night's racing, reducing the amount of available traction drastically. As the track slicks-over drivers have to change how they enter the corners. The idea is to be very smooth with the car, get off the gas earlier and coast longer to let the car roll into the turn. We also have to drive smooth arcs through the corners and be very gentle with the steering. You may have to begin braking much sooner, using less pressure over a longer period of time, to get the car slowed down enough to turn without losing control. We have to prevent upsetting the chassis or breaking the traction of the tires by jerking the steering wheel or stabbing the brakes.

How much brake to use can be difficult to determine and takes considerable practice to learn. Generally speaking, the harder you go into the corner the more brake you will need. As your feeling for what the car is doing improves you will be able to decide on the right amount of brake to keep the car in the groove you want or to set up a pass.

Sometimes when the track is wet with lots of traction we can drive the car very deep into the corner without lifting the throttle, and apply just enough brake to bring the chassis down to help the car take a set. Once the chassis sets, the car turns and you can accelerate out onto the straight without ever getting off the gas. As the track dries out and loses traction race drivers have to get off the gas more and more, as well as increase the amount of braking time.

CORNER ENTRY

When a track stays wet and fast (especially flatter tracks) racers can go into the turns hard and brake hard to bring the back of the car around and set the chassis. These kinds of tracks and conditions may also make the car tight going into the turn. While some drivers will simply drive in harder and brake harder, others compensate by using three-wheel brakes (see sidebar) to help turn the car and get the back end to swing out. Either of these techniques can help make drivers faster, but both can also get drivers in trouble. Driving in too deep and not being able to turn the car before getting to the wall is one obvious problem. Using three-wheel brakes induces a sudden veer to the left that makes applying the right amount of brakes at the right spot very important to prevent spinning out or going through the infield.

I like to run down the straight as close to the wall as I can because that makes the arc into the next corner less severe. Plus, sometimes there is more bite out close to the wall at the end of the straight just after the point where most of the cars are turning into the next corner.

Using the diamond line when the other driver does not expect it can allow you to get through the field. Here Hav-A-Tampa driver Ray Cook (#53) has set the car early and turned under the #3 car to get to a wet spot at the bottom of the track.

We may have to drive a dry-slick track a lot like we do high banks. The best line may be a very wide arc that makes the corner as round as possible. Drivers have to be very smooth with the steering, brakes, and throttle to prevent breaking traction and having to catch the car, all of which slows a car down.

When a track of virtually any design gets dry-slick drivers have to let off the gas farther from the corner to let the car coast farther and slow down more gradually so the tires' contact patches are not over-stressed. We may also have to change the point where we turn in to expand the arc throughout corner. For the same reasons, braking must be gentle and we may "drag" (hold a light pressure on the pedal) the brakes through much of the corner.

MIDDLE OF THE CORNER

On a wet, fast track there is often little or no middle to a corner. Usually we don't have to be very gentle with the gas, braking, or steering because the high levels of traction allow us to be more aggressive. We can drive deep into the turn, and as soon as the car sets, roll back to full throttle and

Hitting the right line through the corner can give you a big enough advantage to complete a pass as you exit the turn onto the straight. Driving in the best line that day could force the other driver to wait a little longer before getting back on the gas.

THROTTLE LINKAGE ADJUSTMENT

Something I have found helpful is making sure the throttle linkage is not only free of binds, but lined up properly. I try to get the throttle lever on the carb and the arm on the gas pedal (see Diagram 1, page 119) at the same angles in the idle position. That way, as I move the gas pedal the carburetor linkage moves through the same arc as the gas pedal arm; so their positions change equally throughout the range of motion. With the arms out of sync the engine may respond suddenly at a certain point of pedal movement. Or worse, one of the arms could go over-center and lock up much more easily. Having both of these arms aligned makes it much easier to control the amount of power being used.

accelerate through the rest of the turn wide open.

On a wet surface the rear of the car can swing out far enough that we have to turn the steering wheel to the right 10 or 15 degrees to "catch" or stabilize the car and prevent it from spinning all the way around. Then as the car accelerates off the corner the steering wheel is gradually returned to center as the car gets back in line heading down the straight. The idea is to get the steering wheel centered at the same time the chassis begins driving straight ahead. If a driver is holding the steering wheel to the right when the chassis neutralizes, the car may veer to the right—sometimes rather hard.

The middle of a dry-slick turn almost always requires being very gentle with everything you do. We can only use as much steering, brake, and throttle as the track will hold and must apply all of them very smoothly. Drivers often have to let the car roll up to or even past the apex of the corner before applying throttle to get off the corner.

The part of this process that seems particularly hard for many new drivers is exiting on time. After they enter the corner, they have trouble waiting until the car is ready to start the exit. If traction is reduced, drivers have to let the car slide until it scrubs off enough speed to allow the chassis to take a set so the car will turn. The car may have to slide what seems like a long time before the chassis hooks up so we can begin accelerating through the remainder of the turn into the line we need to for the best bite.

Getting back on the throttle too early or too hard can produce enough wheel spin to keep a car drifting up to the wall or make the chassis try to take another set. Either way the car slows down (if

it doesn't hit the wall), and the driver of the car following who waited for his car to set before picking up the throttle can drive right past on the low side.

Something I see a lot of drivers, especially new ones, doing is getting on and off the gas repeatedly through the turns. Each time you get on and off the throttle the chassis tries to set again and that delays the car from hooking up. Throttle changes are often needed but they should be small and as smooth as possible.

A lot of drivers think they have to be doing everything really quickly to be fast. Under many conditions, especially dry-slick dirt, drivers have to slow down to be fast. More experienced drivers have learned to wait for the chassis to set before getting back into the throttle so they hit the line they wanted with as much speed as conditions allow.

CORNER EXIT

Unless you are working around traffic, corner exit will always bring you out close to the outside wall. Just like going into the turn, we want to keep the arc from the center of the corner out to the straight as smooth as possible so the rear tires can hold traction and accelerate the car quickest. By the time you get to the middle of the straight you should be up to the outside wall so you can begin setting up for the next turn. Sudden, jerky movements of the steering wheel or throttle can break the tires loose, slowing down the car. If you have to make sudden

Picking just the right spot at which to set the car and turn can take some practice, and even then can get very tricky in traffic. Turn too soon and you wind up going through the infield or the mud, both of which slow you down.

Regardless of what kind of car you drive or what tires you use, keeping the car as straight as possible is almost always faster, especially as the track dries out. The IMCA Modified–style cars are surprisingly similar to Late Models in that they have too much power for their tires to hook to the track. Driver skill is the fastest option in these and all other divisions.

steering corrections to miss traffic you may also have to reduce the throttle to keep control of the back of the car and maintain the most acceleration in a given situation.

I know my car is totally freed up on corner exit if I can feel that little bit of free play in the steering wheel when it is centered. As you gain experience you will be better able to recognize when the car is no longer sliding and is going straight.

USING THE "DIAMOND" LINE

In some situations, primarily on dry flat tracks, running a diamond-shaped line through the corners can be the fastest way around it. Usually the fast line on a high banked track is more of an oval, often a continuously high line through the turns or wherever the best traction is. Using the diamond line on a high bank is seldom fastest.

Imagine four points around the track, one midway down each straight and one in the middle of each turn. The line down the straight should remain smooth and as close to the outside wall as we can, but the middle of the turns will be much sharper than normal. At corner entry we drive low through Turn One and let the car drift up in the center of the corner before braking to set the chassis. Then we turn back down across the track to the bottom of Turn Four and accelerate out onto the straight, drifting back out to the wall again. The car may be slower in the middle of the turn but the lines to that point and back out onto the next straight are much faster than other lines under similar conditions.

Adjusting the Diamond

Even if you don't run 100-lap features like we often do in the Hav-A-Tampa series, your track can change enough in 30 or 50 laps to give you an edge if you can hit the highest traction spots more often than your competition.

One technique to try is moving the point of the

diamond closer to the straight in Turns One or Three to get to a wet spot, usually low in the second or fourth turns. To do this, run into the first part of the turn a little higher, set the car early, and cut back across the track to run through the traction low in Turn Two or Four. Once a driver learns to set the car at a specific spot they can move that set point as needed to get to where the bite is. This technique is also valuable for passing as well as when holes develop or the track tears up. The ability to alter your line through the turns gives a driver more options, more ways to pass, and more ways to be fast.

HUNTING FOR BITE

One of the traits that makes one driver more successful than another is a willingness to search for traction. At Eldora we usually start out running around the wall, but after 20 or 30 laps that line begins to wear out and we still have 70 laps to go. Before long we may be running right around the bottom or cutting across the turns in a diamond fashion to hit a wet spot that helps get out of the turn fast.

You have to constantly hunt for bite because all of the cars out there are wearing out the track. Everyone will probably start the race using the same line, but as that part of the track slows down you have to find where on the track your car will hook up the hardest, and adjust your line accordingly.

On some tracks, depending on what kind of dirt they are made up of, the wet spots may be visibly darker. On other tracks you have to learn to feel for traction—how the tires and chassis react as they slide from slick to areas of bite. Watching other cars can also tell you where the bite is. If they are faster than you (or handling better) in a different line it may be worth your while to move your line and run theirs.

Seat time and driving races is what teaches new drivers how to recognize what the car is telling them. Being able to tell where the car is getting bite is crucial to planning where to let off the gas, turn the car in, and when and how hard to brake. Being overly aggressive and driving in too hard is just as slow as not going in hard enough.

I can't stress enough that learning to drive a dirt car takes time and concentration. Remembering what you did that worked and what didn't work is important. Applying that information to how you set up the car is also important. Be sure that your driving is not what is causing a handling problem before adjusting the chassis, or you could just make matters worse. Be honest with your crew and yourself so you can work together to improve the team's performance. Learning to use everything the track and your chassis setup gives you will bring your lap times down. Control at speed is what allows you to drive around your competition.

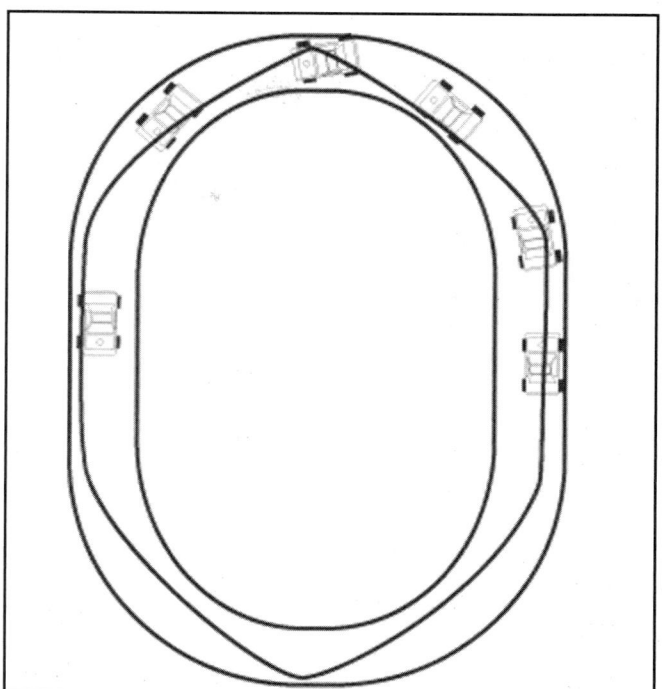

Diagram 2. Notice how a diamond line has a sharp corner at the center of the turn. At this point, the car should be slowed down sufficiently to turn sharply and aimed to drive back down the track, usually across the bottom of the corner exit where there is often more bite.

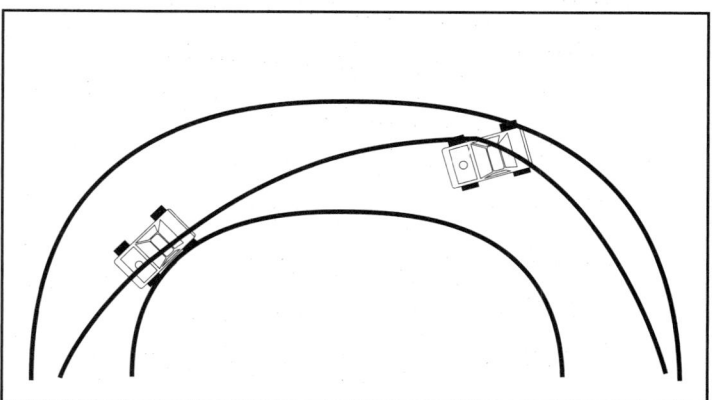

Diagram 3. Sometimes the apex of the turn is moved closer to the entrance to allow you to set the car early and turn under another car. This works especially well when there is good bite low in the exit corner. While the car you are passing continues to the nomral apex, you are already set, turned and accelerating down the track to try and be the first to the straightaway.

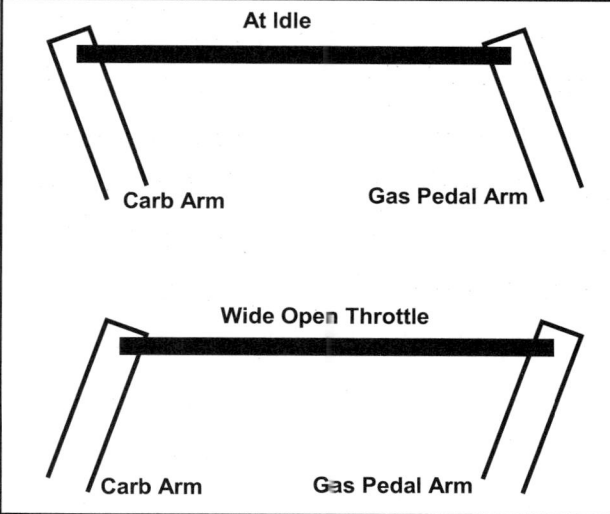

Diagram 1. Having both the carb's throttle lever and the lever of the gas pedal at the same angles helps make the throttle feel smoother and more predictable. In some cases, misalignment of these arms may contribute to stuck throttles.

NOTES

GENERAL MOTORS

Big-Block Chevy Engine Buildups: 1-55788-484-6/HP1484
Big-Block Chevy Performance: 1-55788-216-9/HP1216
Camaro Performance: 1-55788-057-3/HP1057
Camaro Restoration Handbook ('67–'81): 0-89586-375-8/HP1375
Chevelle/El Camino Handbook: 1-55788-428-5/HP1428
Chevy S-10/GMC S-15 Handbook: 1-55788-353-X/HP1353
Chevy Trucks: 1-55788-340-8/HP1340
How to Hot Rod Big-Block Chevys: 0-912656-04-2/HP104
How to Hot Rod Small-Block Chevys: 0-912656-06-9/HP106
How to Rebuild Small-Block Chevy LT-1/LT-4: 1-55788-393-9/HP1393
John Lingenfelter: Modify Small-Block Chevy: 1-55788-238-X/HP1238
LS1/LS6 Small-Block Chevy Performance: 1-55788-407-2/HP1407
Powerglide Transmission Handbook:1-55788-355-6/HP1355
Rebuild Big-Block Chevy Engines: 0-89586-175-5/HP1175
Rebuild Gen V/Gen VI Big-Block Chevy: 1-55788-357-2/HP1357
Rebuild Small-Block Chevy Engines: 1-55788-029-8/HP1029
Small-Block Chevy Engine Buildups: 1-55788-400-5/HP1400
Small-Block Chevy Performance: 1-55788-253-3/HP1253
Turbo Hydramatic 350 Handbook: 0-89586-051-1/HP1051

FORD

Ford Windsor Small-Block Performance: 1-55788-323-8/HP1323
Mustang Performance (Engines): 1-55788-193-6/HP1193
Mustang Performance 2 (Chassis): 1-55788-202-9/HP1202
Mustang Performance Engine Tuning: 1-55788-387-4/HP1387
Mustang Restoration Handbook ('64–'70): 0-89586-402-9/HP1402
Rebuild Big-Block Ford Engines: 0-89586-070-8/HP1070
Rebuild Ford V-8 Engines: 0-89586-036-8/HP1036
Rebuild Small-Block Ford Engines: 0-912656-89-1/HP189

MOPAR

Big-Block Mopar Performance: 1-55788-302-5/HP1302
How to Hot Rod Small-Block Mopar Engine Revised: 1-55788-405-6/HP1405
How to Maintain & Repair Your Jeep: 1-55788-371-8/HP1371
How to Modify Your Jeep Chassis/Suspension for
  Offroad: 1-55788-424/HP1424
How to Modify Your Mopar Magnum V8: 1-55788-473-0/HP1473
How to Rebuild Your Mopar Magnum V8: 1-55788-431-5/HP1431
Rebuild Big-Block Mopar Engines: 1-55788-190-1/HP1190
Rebuild Small-Block Mopar Engines: 0-89586-128-3/HP1128
Torqueflite A-727 Transmission Handbook: 1-55788-399-8/HP1399

IMPORTS

Baja Bugs & Buggies: 0-89586-186-0/HP1186
Honda/Acura Engine Performance: 1-55788-384-X/HP1384
Honda/Acura Performance: 1-55788-384-X/HP1384
How to Hot Rod VW Engines: 0-912656-03-4/HP103
Mitsubishi & Diamond Star Performance Tuning:
  978-1-55788-496-1/HP496
Porsche 911 Performance: 1-55788-489-7/HP489
Rebuild Air-Cooled VW Engines: 0-89586-225-5/HP1225
The VW Beetle: History of The World's Most Popular Car:
  1-55788-42 '21
The Volks   eetle Handbook: 1-55788-483-8/HP1483

HANDBOOKS

Automotive Detailing: 1-55788-288-6/HP1288
Auto Electrical Handbook: 0-89586-238-7/HP1238
Auto Math Handbook: 1-55788-020-4/HP1020
Automotive Paint Handbook: 1-55788-291-6/HP1291
Auto Upholstery & Interiors: 1-55788-265-7/HP1265
Car Builder's Handbook: 1-55788-278-9/HP1278
Classic Car Restorer's Handbook: 1-55788-194-4/HP1194
Engine Builder's Handbook: 1-55788-245-2/HP1245
Engine Cooling Systems: 978-1-55788-425-1/HP1425
Fiberglass & Other Composite Materials Rev.: 1-55788-498-6/HP1498
The Lowrider's Handbook: 1-55788-383-1/HP1383
Metal Fabricator's Handbook: 0-89586-870-9/HP1870
1001 High Performance Tech Tips: 1-55788-199-5/HP1199
1001 MORE High Performance Tech Tips: 1-55788-429-3/HP1429
Paint & Body Handbook: 1-55788-082-4/HP1082
Performance Ignition Systems: 1-55788-306-8/HP1306
Pro Paint & Body: 1-55788-394-7/HP1394
Sheet Metal Handbook: 0-89586-757-5/HP1757
Welder's Handbook Revised: 978-1-55788-513-5/HP1513

INDUCTION

Holley 4150: 0-89586-047-3/HP1047
Holley Carbs, Manifolds & F.I.: 1-55788-052-2/HP1052
Rochester Carburetors: 0-89586-301-4/HP1301
Tuning Accel/DFI 6.0 Programmable F.I.: 1-55788-413-7/HP1413
Turbochargers: 0-89586-135-6/HP1135
Street Turbocharging: 1-55788-488-9/HP1488
Weber Carburetors: 0-89586-377-4/HP1377

RACING & CHASSIS

Bracket Racing: 1-55788-266-5/HP1266
Chassis Engineering: 1-55788-055-7/HP1055
4Wheel & Off-Road's Chassis & Suspension: 1-55788-406-4/HP1406
Dirt Track Chassis & Suspension: 978-1-55788-511-1/HP1511
How to Make Your Car Handle: 0-912656-46-8/HP146
How to Get Started in Stock Car Racing: 1-55788-468-4/HP1468
How to Build a Winning Drag Race Chassis & Suspension:
  978-1-55788-462-6/HP1462
Racing Engine Builder's Handbook: 1-55788-492-7/HP1492
Stock Car Racing Engine Technology, 978-1-55788-506-7/HP1506
Stock Car Setup Secrets: 1-55788-401-3/HP1401

STREET RODS

How to Build a 1934–'35 Chevy Street Rod: 978-1-55788-514-1/HP1514
How to Build a 1935–'40 Ford Street Rod: 1-55788-493-5/HP1493
Street Rodder magazine's Chassis & Suspension Handbook:
  1-55788-346-7/HP1346
Street Rodder's Handbook, Rev.: 1-55788-409-9/HP1409
Street Rodding Tips & Techniques, 978-155788-515-9/HP1515